History of Life Sciences

and Medicine 1

SCIENCES

History of Sciences, Field Director – Jean-Claude Dupont

Life Sciences, Subject Head – Jean-Claude Dupont

History of Life Sciences and Medicine 1

Antiquity and the Middle Ages

Coordinated by
Lorenzo Perilli

iSTE

WILEY

First published 2026 in Great Britain and the United States by ISTE Ltd and John Wiley & Sons, Inc.

ISTE Ltd
27-37 St George's Road
London SW19 4EU
UK

www.iste.co.uk

John Wiley & Sons, Inc.
111 River Street
Hoboken, NJ 07030
USA

www.wiley.com

Library of Congress Control Number: 2026931542

British Library Cataloguing-in-Publication Data
A CIP record for this book is available from the British Library
ISBN 978-1-78945-208-2

ERC code:
SH6 The Study of the Human Past
SH6_14 History of science and technologies, environmental history

The manufacturer's authorized representative according to the EU General Product Safety Regulation is Wiley-VCH GmbH, Boschstr. 12, 69469 Weinheim, Germany, e-mail: Product_Safety@wiley.com.

Contents

Presentation of Field “History of Science”

Jean-Claude DUPONT
CHSSC, Université de Picardie Jules Verne, Amiens, France

The “History of Science” field of the *SCIENCES* collection is closely related to the sections devoted to scientific and technical disciplines, on the one hand, and to epistemology, on the other hand. In addition to providing a summary of current methods and issues in the history of science, all of the volumes offer an overview on the evolution of knowledge in mathematics, the physical and natural sciences, medicine, the humanities and technology, from the beginning to the present day, as well as in different cultural areas. This is aimed at non-specialists and scientists alike, providing them with a useful perspective on the origins and foundations of scientific knowledge, while keeping up with contemporary research in the history of science.

– Field Director: Jean-Claude Dupont.

– Subject Heads of the English edition:

- “Issues, General Themes and Methods of History of Sciences”: Stéphane Tirard.

- “Logic and Mathematics”: Sabine Rommevaux-Tani.

- “Astronomy, Physics, Chemistry and Earth Sciences”: Christian Bracco.

- “Life Sciences”: Jean-Claude Dupont.

- “Humanities and Social Sciences”: Alexandre Escudier.

History of Life Sciences and Medicine 1,
coordinated by Lorenzo PERILLI. © ISTE Ltd 2026.

Medicine and the Life Sciences in the Ancient World: An Introduction

Lorenzo PERILLI
Università degli Studi di Roma Tor Vergata, Rome, Italy

Medicine is an art of knowledge as well as of action – a *techne* par excellence. It demands the cunning of the eye, hand and brain. It has accompanied the history of humankind since its very origin: the bite of a wild animal, a broken leg, the consequences of a fight – all required intervention. Medicine was different from all other disciplines because of the testing ground for its theories and practices: life or death.

The earliest written text of Western civilization, the *Iliad,* collects stories that for centuries had been circulated by word-of-mouth and fixes them in writing. Here, doctors emerge from among the Greek warriors campaigning at Troy, but medical practice is also the task of gods and heroes alike. Homer, the figure to whom the composition of such an extraordinary work was later attributed in conjunction with the *Odyssey*, exhibited curiosity, familiarity and even competence in the fields of anatomy, physiology and treatment. He describes wounds, bandages, drugs and surgery across more than 150 instances. The medical scenes are so many and detailed, their technical accuracy so precise, that a scholar once remarked, with a fair degree of imagination (and outdated perspective), that the poet must have been an army doctor who accompanied the Greek warriors.

History of Life Sciences and Medicine 1,
coordinated by Lorenzo PERILLI.

This was not the case. Medical knowledge and curiosity, as exemplified in the earliest poems, were common heritage. They accompanied not just Greek culture but every culture, and each culture approached this knowledge, and therefore the human body and health, in its own peculiar manner, with an approach that reflected the most general vision of humanity, of the world, and of knowledge.

It all revolves around the very notion of medicine. It is difficult to comprehend the concept that the ancients had of medicine if we proceed with the categories of modern medicine, just as it is difficult, and misleading, to look at ancient science through the lenses of modern science. The two are not equivalent: nor should the former be regarded as preparation or anticipation of the latter.

Ancient medicine has often been attributed a system of ideas on the basis of what we term the "paradigm of rationality" – the paradigm that rendered modern Western science the dominant system. However, in ancient times, medical theory and practice actually possessed a multifaceted nature, in which philosophy, faith and a form of anthropology linked to the deep-rooted religious dimension – not devoid of elements of superstition – intersected. Shrines dedicated to the god of medicine Asclepius (named Aesculapius in Rome) and his father Apollo could be found in every city of the Mediterranean. They hosted doctors and priests, maintained medical writings and records, and served as a meeting point for hundreds of faithful patients who came seeking care for both the soul and the body, or for young people eager to learn, intellectuals and Roman emperors seeking reassurance or treatment, even if only for their hypochondria.

Medicine was a collective and sociopolitical phenomenon. It did not live in isolation but made use of skills and experiences accumulated over time and across regions. It benefited from the traditions of ancient civilizations, especially Egyptians, and exploited the practices of many practitioners in cities and sanctuaries, sometimes incorporating even magical components. As the Greek historian Herodotus stated, the Babylonians could do without doctors: among them, he says, the ill were brought to a public space where anyone who had experienced the same illness before could make suggestions, and indeed passersby were obliged to do so (Herodotus, *Histories* I 197).

The Greeks viewed it differently. For them, as Homer noted, "a doctor is worth many men", and doctors were appointed by cities to serve the health of all – rich and poor alike – considering their work a mission of life. This mission was later formalized in the *Hippocratic Oath* (which by the way was not by Hippocrates): to the extent that, as stated in a famous 304–303 BCE inscription in the sanctuary of Asclepius at Athens, "Phidias the doctor has always acted for the benefit of the

Athenian people and generously cared for the Athenians who needed it, and offered to serve as a public doctor free of charge, thus demonstrating his benevolence toward the city". For this purpose, the doctor Phidias of Rhodes received public honors and a commemorative stone inscription at state expense.

Egypt itself provided the closest term of comparison, and indeed a benchmark, for the Greek world. The Greeks themselves acknowledged it: for Homer, Egyptian doctors were "the most expert of all men" (*Odyssey* IV 231f.); Herodotus repeatedly recalled with admiration the presence of Egyptian doctors in Greece; and later, the philosopher and Christian theologian Clement of Alexandria listed the following six books, which, according to him, contained the science of Egyptian doctors: *On the Structure of the Body, On Diseases, About Instruments, About Drugs, On the Eyes* and *On Women's Diseases*.

Mesopotamia, China and later India and the Near East of Arab-Syriac mediation constitute the pieces that help complete the mosaic of ancient medicine: the Western, and thus Greek, paradigm has proven successful throughout history, but this paradigm and its path would not be understood without comparison to different models. Comparison allows us to see aspects that, with an "isolationist" perspective, risk going unnoticed. The aim is not, or no longer, to identify alleged "superiorities" but rather to understand how much the context, the *Sitz im Leben*, the value system and state of knowledge affect medicine and science and, conversely, how much the latter inform the former.

The Greek rational paradigm was the ultimate victor. The entire history, not just of medicine but also of science and knowledge, is shaped by this legacy, our key to the interpretation of the world; we are the heirs of this paradigm, with its merits and limitations. This paradigm, born with the Achilles of the *Iliad* and cemented in the 5th century BCE, was also the natural consequence of the theoretical reflection that Greek medicine undertook at an early stage about itself and its modus operandi. This above all distinguished it from other disciplines: its conceptual and theoretical legacy remains unique, as does its self-awareness, which is evident both in numerous surviving medical works and sources and in other testimonies. Among the latter is Plato's famous argument: there are, said the philosopher, two types of doctor. One learns by experience, through passions, and can only cure those alike enslaved in a like manner by passions; the other, the doctor par excellence, a doctor resembling Hippocrates, knows the theoretical foundations of his art, can "provide a reason" (*logos*) and can assign a rational basis to his actions. He is a free man and will cure the free.

This vision found fertile ground in Greek thought, met Aristotle and Diocles of Carystus, nourished Hellenistic medicine, informed the debates between opposing schools such as the empirical and methodical, and reached its height six centuries later in the "new Hippocrates": Galen. At this point in history, in 2nd century Rome, medicine as a rational science and its relationship with philosophy were fulfilled, accompanied by an extraordinary level of technical knowledge, laying the foundations for European medicine until the modern period. Only starting in the 16th century, with Andreas Vesalius' contribution to anatomy, did the path begin that would overcome Galenic medicine, taking two more centuries to complete and enter the new world.

The "other" medicine, that of the healing cults spreading all over the Greco-Roman world, continued competing, if less visibly as less institutionalized. Creeping into the transformations of medieval society, the dimension of faith inherent in care and healing constantly accompanied it, at times becoming dominant. There were many doctors, methods, "protocols", but if observed backward from our contemporary vantage point, the non-Hippocratic approach was disfavored precisely for being non-Hippocratic, that is, for remaining outside of the dominant paradigm.

The scope of "medicine" in the ancient world went far beyond that of a "science or practice of the diagnosis, treatment, and prevention of disease", as modern dictionaries have gone. It was a science of man and the living, a science of the relationship to the world and to others. Understanding this helps us appreciate the importance of comparison with different cultures, which, while possibly similar nosologically, were not necessarily comparable from a diagnostic or therapeutic point of view, especially with respect to the very idea of man. The medicine of China or India can reveal much more about conceptions of the body and its interrelation with the external, sub- and supralunar-world and even about theories on politics and state organization than profound theoretical speculations. Thus, we Westerners learn to look at our neighbors' complexity with a different eye.

Today's perspective on ancient medicine has changed remarkably in just half a century. It has left the narrow circles of disciplinary historians and insightful philologists of the late 19th and early 20th centuries, revealing its many facets and its contributions to the understanding of history, society, religion, anthropology, economics and, of course, "science" in ancient societies. Returning medicine to its place in history and context enabled different readings and reinterpretations of the abundant evidence available. This encompasses written sources – medical and otherwise, including numerous inscriptions – as well as archaeological, figurative and numismatic evidence.

Medicine found an indispensable tool in writing, giving us for it alone among all ancient *technai* a substantial corpus of works, starting from the high point of ancient Western culture: 5th century BCE Greece, the age of the Parthenon and the art of Phidias and Polykleitos; Parmenides, Socrates and philosophy; Sophocles and tragedy, Aristophanes and comedy; Thucydides and historiography; Hippocrates of Chios and mathematics; and Hippocrates of Cos and medicine. The formalization of medicine and life sciences is among the richest moments in Western history.

The main complete written sources still available today (thus excluding fragmentary/indirect evidence) are concentrated – by a coincidence of fate, or for some precise reason – around the 5th–4th centuries BCE, and then again in the 1st–2nd centuries CE when Galen concluded the era of creativity, giving way to compilers and erudition that led to the Middle Ages. The mediation of the Arab world and the Near East that took place in the 6th–9th centuries CE was essential, translating Greek scientific texts into Syriac and Arabic, while also contributing original thoughts, and represented a vital phase of the relocation, circulation and reformulation of knowledge and ideas.

This new perspective on ancient medicine has led to a more nuanced understanding of disease; doctors' training, role and social function; and doctor–patient relationships. It has led to more precise definitions of pathologies that, like mental illnesses, were not always seen as such, while also examining individual and societal reactions to infectious epidemics.

Furthermore, awareness of the usefulness of placing the Greek world in a broader context came, and notions of "miraculous uniqueness" as fossilized by the epigones of Neoclassicism were discontinued, whereas surrounding cultures, even as distant as China's, were the object of beneficial comparison.

Broader and more structured attention and better appreciation granted ancient medicine a growing interest and commitment. Publishing initiatives, research centers and national and international projects have provided scholars and the interested layman with reliable editions of texts and dependable wide-ranging studies, facilitating new discoveries and fostering scientific cooperation and dialog between specialists from different fields. Much remains to be done – the field lies open.

1

Egyptian Medicine

Paola COSMACINI[1] **and Christian ORSENIGO**[2]
[1]*Independent Researcher, Rome, Italy*
[2]*Museo Civico di Crema e del Cremasco, Italy*

1.1. Medical thought and the *Weltanschauung* of the Egyptian man

The assimilation – and even before that, the understanding – of medical thought as far back in time as that of ancient Egypt is a complex case. The geographical context and the vast historical period involved, as well as the cultural framework resulting from the revolution in medical thought in the 17th century, prevent us from taking a deep dive into the reality of ancient Egypt, whose population is assumed to have grown from around 350,000 individuals in predynastic times to 5 million around 150 BCE (Butzer 1976, p. 83). Importantly, although certain formulas and gestures are essentially unchanged in the texts in their reproduction and transmission and attest to the presence of a therapeutic ritual consolidated over the centuries as part of an evolutionary or involutionary process (Ritner 2000, pp. 107–108), we cannot speak of a real corpus of medical literature, since we only possess random relics, so much so that "our material is very scanty and we may have to draw far-reaching conclusions from less than a dozen texts" (Sigerist 1951, p. 298). With such a small number of texts available over a period of more than two millennia, it is not possible to approach medical literature synchronically by grading its possible historical progression (or regression). In a similar manner, human remains (skeletons and mummies), as numerous as they are, can rarely be statistically representative of an entire population and its diseases.

It is necessary to use different reading keys so that medical thought be never separated from the other cultural events of the Egyptian people and above all from

History of Life Sciences and Medicine 1,
coordinated by Lorenzo PERILLI. © ISTE Ltd 2026.

their conception of the world, just as, in turn, "the world, like a gigantic hieroglyphic text, had numerous and not always obvious levels of reading for the Egyptians" (Betrò 2005, p. 107). The "magico-medical" reading key is one of these (Ritner 2001). While, in a certain sense, it remains a valid consideration that "if medicine stems from magic, magic is only one aspect of religion, and all three, among the Egyptians, have always remained intimately intertwined" (Lefebvre 1956, p. 9); Egyptians "were the first in history to dare look across the abyss that separates magic from science" (Ghalioungui 1963, p. 171). Indeed, advances in research suggest that "the prevalence and influence of 'irrational' treatments were probably less important than previously thought" (David 2004, 2008b).

The supernatural forces represented by religion and magic naturally found their place in the medical act of treatment and therapy, since medical thought was also coherently integrated into the Egyptian understanding of reality and justified by a harmonious and global vision of the unchanging cosmic order, conceived through the notion of "Maat" (Assmann 1990, p. 15–39). Consequently, in Egyptian medicine, as in rational practice, gestures, the religious aspect and the magical approach could only coexist, materializing in an intermediary expression between the sacredness of ritual and the practice of craft. Thus, in the etiopathogenesis of diseases, rational and irrational motivations interweave with one another, helping to fill their respective disparities, whether they be tangible or of a divine, magical, mysterious and occult nature, such as the vengeance of a god, a demon, a dead person or an enemy (Lucarelli 2017). This is all the more evident, as "once we leave the realm of trauma, ancient Egyptian medicine loses much of its pragmatism and there is increasing evidence of empiricism, speculation, unproven remedies and magic" (Nunn 1996, p. 59).

This is a concept that must be part of an instinctive, "natural" approach, which we define today as supernatural or magico-religious, the result of which is not the simple addition of different factors but a reciprocal strengthening that reinforces both the objectivity of the ailment and the subjectivity of the sickness, perfectly framed in the *Weltanschauung* of Egyptian man and closely linked to his profound moral sense. This is also why, in ancient Egypt, the object of medicine was man in his entirety.

1.2. The observed body

For Egyptians, the purpose of embalming was to prevent the dead from dying: the condition for survival was for the body to retain its physical characteristics, in other words, to be recognizable (see, for example, the *Book of the Dead*, Chapter 154, "Spell for not letting the corpse perish"). The practice of mummification ennobled bodily integrity, from the *khat* state to the *sah* state, that is, the incorruptible body of man purified of its perishable parts: man not only "had a body" but, at the same time, "was a body", and bodily integrity after death and its recognizability by the incorporeal part of the individual were essential conditions of immortality. Man was considered a unitary entity, and contrary to what Herodotus transmitted, Egyptian thought did not include concepts such as transmigration of the soul or metempsychosis. Thus, it is clear that "the culture of ancient Egypt offers rich resources for analyzing the Egyptians' conceptualization of the body and the embodied self" (Riggs 2010, p. 1–2).

Given that the body was the "subject", and that one of the sacred and secret books of medicine, which has never been preserved, is said to have dealt with "the structure of the body", it is not surprising to find that it was certainly the "object" of much thought, reflected both in the meanings given to the word "body" and in the multiplicity of hieroglyphic signs expressing the various parts of the body (Walker 1996, pp. 265–279), as well as in its iconography. However, we cannot avoid mentioning the obvious idealization of body rendering in Egyptian art because:

> In general, anatomical observation is subject to the idealizing function: the two fundamental models of position and proportion do not operate to imitate the visible appearance but to affirm the ideal reality, not to record the mutable but to establish the immutable. The human figure, therefore, is not so much represented according to appearance as constructed, or reconstructed, according to stable laws and on the basis of ideal principles (Moscati 1975, p. 51).

Nevertheless, since the Old Kingdom, the bodies of men and women have been, if not dissected, at least studied in terms of their physiological normality and, occasionally, their pathological anomaly (see, for example, Filer 1996, pp. 29–36). In the figurative field, by way of example, let us recall the evidence of sculptures of achondroplastic dwarfs (such as the family group of Seneb the dwarf) as early as the 4th–5th dynasties, such as the "famine" scene found on a limestone block in the processional causeway of the pyramid of Unas at Saqqara (5th Dynasty), the gynecomastic obesity of Ankhmahor represented on a bas-relief (6th Dynasty), Bak's bulging abdomen on his stela (18th Dynasty), Roma's crippled leg resting on a stick as depicted on his stela (18th–19th dynasties), various representations of umbilical and scrotal hernias from the 6th Dynasty (tomb of Ptahhotep) to the New Kingdom (tomb TT100 of Rekhmire, and the "elephantine" figure of the Queen of Punt on a bas-relief from the temple of Hatshepsut at Deir el-Bahari (18th Dynasty). Certain scenes from the mural decoration of the tomb of the sculptor Ipuy (TT217, 20th Dynasty) at Deir el-Medina, where the gardener appears kyphotic, document numerous accidents at work by craftsmen in the Valley of the Kings: in this case, it is everyday pathology that is "photographed" and "the presence of individuals with physical anomalies seems to be no more than a reflection of everyday activity" (Robinson 2017, p. 28). On the other hand, the particular representation of the pharaoh Amenhotep IV/Akhenaten, which perhaps expresses an iconographic convention that intends, among other things, to concretize the concept of the "father and mother" of all humanity, deserves to be discussed separately.

From a strictly medical point of view, we do not know whether Egyptians also had a global vision of the body; in other words, whether they viewed the body as a unitary structure and a simple aggregate of parts, that is, a "dispersed body". It was certainly not divided "functionally" but "anatomically" into independent parts, each of which was likened to a deity (Duquesne 2002). While references to this concept are not lacking in the *Book of the Dead* (for example, Chapter 42), they are particularly numerous in the *Coffin Texts* (see spell 761: "Gathering the members of a spirit for him in the realm of the dead") (Nyord 2009). In practice, Egyptians "saw a 'marionette', an aggregated multitude of individual members, not an organic whole controlled from the center outwards" (Assmann 2005, p. 26); moreover, these elements were "the plaything of higher forces, beneficial or harmful, which the Egyptians would represent, very materially, by active breaths penetrating the body" (Bardinet 1995, p. 60). This concept, which is well understood if placed within a broader context of Egyptian thought that considers the most ancient and primitive forms of conceptualization, is better understood if related to the meaning given to the "tubular" structures of the body (*metw*), to the concept of the blood "that circulates through the body and makes it whole and intact" (Riggs 2010, p. 2), and to the concept of the heart as the center of life and death.

Observation of the body, dead or alive, is, in any case, very scrupulous. From the carefully described traumatology often observed during combat and accidents in peacetime, we can acquire knowledge of the different locations of the body, with the relative arrangement of anatomical structures. The study of the body was carried out in vivo, without any postmortem objectification; however, there is no doubt that the link with insight into the art of mummification raised the level of knowledge. Indeed, even though dissection was forbidden, Egyptians opened bodies for over 2,000 years, and the extent of anatomical knowledge that resulted from this can be seen both in the richness of the vocabulary – over a hundred anatomical terms – and in the precision of the descriptions of certain body parts, suggesting that an "anatomical view" already existed (Dawson 1942, p. 188; Vycichl 1972). The link – or at least the dialogue – between physicians and embalmers is in fact now proven, not just because the word *swnw* also meant *to embalm* (Nunn 1996, p. 115) from the 27th Dynasty onward, but above all because it was present in everyday life – the linen strips used for medicine being the same as those used for embalming (P. Edwin Smith, Case 9, Gloss. A, V, 4-5: (Breasted 1930, p. 224)) – or in the family environment itself. Physicians either had an *Embalmer's manual* at their disposal – the trace of which had been lost but which might have contained anatomical details (P. Edwin Smith, Case 19, Gloss. A VII, 19–22: (Breasted 1930, p. 281)) – or a small "manual on embalming" contained in a papyrus of magico-medical material, specifically written to instruct court physicians (P. Louvre E32847 to 16, 17–19.4 (Bardinet 2018, pp. 211–226)).

Talking about anatomical observation is an opportunity to reference the foundation of Egyptian medicine. For a long time, the process of embalming and the management of trauma were considered practices underlying the origin of medical art in the Nile Valley. However, since the early 1990s, an alternative approach has been proposed, which is now widely accepted and considers the maintenance of animal health and ritual sacrifice as the origin of Egyptian medicine (Gordon and Schwabe 2004): a less anthropocentric and more zoocentric viewpoint, in line with Sigerist's idea that "the main sources of anatomical knowledge were the kitchen and the cult" (Sigerist 1951, p. 352). Consequently, the preconceived idea that the first form of anatomical observation by Egyptians concerned humans, rather than animals, can be put into perspective on the basis of certain considerations, such as the antiquity of the hieroglyphs used to indicate most human organs, which are the graphic transposition of indisputably animal organs (for example, Ghalioungui 1963, pp. 68–69). Moreover, it should be remembered that the written codification of the Egyptian language in the form of hieroglyphs occurred at a time when embalming, and evisceration before it, was not yet practiced, even though medical thought had clearly already taken its first steps. These hieroglyphs indicate that Egyptians' knowledge of the internal structure of animals would have preceded that of the human body. The processes of contamination between human and veterinary medicine and their subsequent autonomous development are still a matter of debate among specialists, who alternate between potential positions when they assert that comparative medicine "has probably been fundamental to medical progress since the dawn of medical science" (for example, Schwabe 1986, p. 145) and more conservative positions.

That being said, it is no coincidence that the only known text devoted to veterinary practice, dating from the end of the 12th Dynasty/beginning of the 13th Dynasty (c. 1850–1700 BCE), was coeval with medical papyri, testifying to very early attention to the problem. The text is known as the Kahun veterinary papyrus (UC32036 lot LV.2; plate 4 (Griffith 1898, pp. 12–14; Collier and Quirke 2004, p. 55)) and deals with the care of animals, particularly cattle: this was knowledge deemed so important that it had to be transmitted in writing. The veterinary papyrus follows a logical progression of treatments, which brings it close to the Edwin Smith papyrus (Lord 2011). In addition, references to veterinary knowledge can also be found in a section of the Rammeseum Onomasticon (P. Ramesseum D/P. Berlin 10495), also found with medical papyri.

In addition, as far as ritual sacrifice was concerned, even if it was more linked to the religious and magico-cultural sphere than to that of everyday practice, the observations that priests could draw from the dissection of animals certainly led to a rudimentary knowledge of animal anatomy and physiology, which was subsequently transposed to that of the human body. In the apparent absence of a specific term to describe the veterinarian, it has been suggested that *physicians* or *priests of Sekhmet* might have been used, although it seems that their intervention was more focused on the suitability of livestock for sacrifice (contra (Lord 2016, p. 146)).

1.3. Healers

The ancient Egyptian term for physician is *swnw*. The word, which seems to have been attested as early as the 1st Dynasty, is expressed in hieroglyphs by the triliteral sign *swn* (representing an arrow) accompanied by the biliteral sign *nw* (representing a bowl). The interpretation that links the arrow hieroglyph to the primary purpose of medical art (the healing of wounds or disease) seems to be linked to its impact on the body or to the disease itself, which is as swift and damaging as an arrow (Étienne 2016, p. 255). The word is linguistically close to "pain" (*swnyt*) and "suffering" (*swn*): it would therefore designate someone who has to deal with pain and suffering or, more precisely, "one who cares for those in pain and suffering" ((Jonckheere 1951a, pp. 28–45; Contin 2005, p. 116); contra (Jean 2012, p. 34)).

It is likely that the training of physicians first took place within the family, where the techno-practical heritage could also be passed on secretly, and then continued in the field of self-taught experience and within the "houses of life", centers of excellence with a considerable role within Egyptian society, where study was based on verbal tradition and written concepts (Eyre 2013, pp. 311–315).

Within society, there was a hierarchical structure that ranged from the great physician (*wr*) to the simple physician through characters identified as inspector (*sehedj*), overseer (*imy-er*), controller (*kherep*) and chief (*hery*). The names of approximately 150 physicians are known from the Old Kingdom to the 27th Dynasty (Nunn 1996, pp. 211–214), including one female figure: Peseshet. This woman is the owner of a stela in the mastaba of Akhet-hetep (G8942), in which she is referred to as the "overseer of female physicians" (Ghalioungui 1975; Cole 1987).

During Herodotus' visit to Egypt – which took place after the reform commissioned by the Persian central authority to improve the offices of the "house of life" and was carried out by Udjahorresnet from Sais (Lopez 2015, pp. 294–298) – he emphasized the degree of specialization he found, which is confirmed by the terms, usually of body parts, that follow the word *swnw* on many monuments, such as "doctor of the eyes" (*swnw jrty*), "doctor of the belly" (*swnw khet*), "interpreter of the liquids hidden in the *netjet*" (*swnw Aaa mw em khenw netjet*) or other figures such as the "shepherd of the anus" (*nerw pekhwt*) or the "dentist" (*jbhy or jry jbeh*) (Wugh 1995, I, pp. 300–302). While this was true in 430 BCE and especially in the Old Kingdom, the same cannot be said of the Middle and New Kingdoms (Nunn 1996, p. 119, Plate 6.3).

Interestingly, "there is also evidence of 'local healing' in the New Kingdom and later on, such as the Deir el-Medina workers who practiced healing techniques, and the 'wise women' (*rekhwt*) who were skilled at detecting and combating spirit possession" (Forshaw 2014a, p. 32). In this respect, the discovery at Deir el-Medina itself of a tattooed female mummy who may have been a "wise woman" (*rekhyt*: literally "the knowing" or "the learned") or, perhaps, some sort of "scorpion charmer" (*kherep Serket*) is remarkable (Austin and Gobeil 2017).

The therapeutic act involved the execution of certain maneuvers, the administration of medicines, but also the formulation of spells and the performance of religious practices. It was not practiced by a single category of individual, but by various types of "healer" who, from time to time, took part in the therapeutic act itself: thus, in addition to the physician, there was the magician (*saw*, literally "the protector") (among others (Lucarelli 2020)), as well as various categories of priests, such as the "lector priest" (*khery-heb*, literally "he who carries the ritual book" (Forshaw 2014b, pp. 115–121), the simple "*wab*-priest" (*wab*) and the "*wab*-priest of Sekhmet" (*wab Sekhmet*) (see, for example, von Känel 1984). The latter is the devout priest of the powerful goddess, the giver of life, prosperity and health, whose power, once penetrated into the body, can heal and protect against disease. At the same time, Sekhmet was linked to epidemics and "disease-demons", her messengers. Her power was so great that she feared that the intervention of her priests was indispensable; therefore, she acquired healing abilities that were considered equal to or almost equal to those of the *swnw*. In addition, the aforementioned "scorpion charmer" was responsible for the treatment of scorpion stings and snake bites, as shown in the treatise on ophiology (P. Brooklyn 47.218.48 and 47.218.85), in which the systematic description of the pathology (and treatment) of ophidian bites is characterized by a rational and pragmatic methodological approach, although not without its magical elements (Sauneron 1989).

This can be seen as early as the Old Kingdom, with the emergence of a differentiation of roles assumed by the various "healers", who, however, especially in the Old Kingdom and the First Intermediate Period, were able to focus on their own. In this case, they bore different titles and fulfilled different functions, such as "the best therapeutic practice combined the technical skills of the physician with the bedside manner and faith healing of the priest and magician" (Wilson 1964, p. 1), such as a figure like Hery-shef-nakht (12th Dynasty), "chief of the king's physicians", who held all three titles, being *wr swnw*, *wab*-priest of Sekhmet and overseer of the magicians (*imy-er hekaw*) (Anthes 1928, pp. 33–35, pl. 19). Two famous passages confirm that *swnw*, *wab* and *saw* may have had interchangeable roles in the clinical examination of the person and were called upon to determine the state of the suffering individual (P. Edwin Smith, Case I, Gloss. A, and Ebers #854a).

Thus, given that reason, religion and magic most often coexisted in the act of healing and that great hope was placed in their mutual synergy and power, we must conclude that "in ancient Egypt, the healer was physician, priest and magician all in one, representing science, religion and magic. The subject of medicine is the human being as a whole"[1]. On the other hand, this life-saving interaction between the actors in the healing process indirectly shows that the powers of the *swnw* were limited and thus lays the foundations for the idea that "the physician's power" could only come from a "profession".

1.4. Written sources

According to Clement of Alexandria, the Egyptians possessed six sacred and secret books on medicine, which he himself had seen carried in ritual processions, along with the "philosophical" books. Among these papyrus scrolls on medical subjects, one dealt with the structure of the body, one with diseases, one with instruments, one with remedies and the last two with eyes and women (Stromata, book VI, 4, 35.1–37, 37.3). This source not only testifies to the existence of a corpus of medical literature at the time but also to the fact that medicine was already considered an autonomous and articulated form of knowledge (Dolzani 1993, p. 107; Ritner 1992).

1 See: www.ub.uni-leipzig.de:9000/54d8b146f3e91b134c000038/apps/55327c81569c2c2d3d000ed3/en.html.

The study of ancient Egyptian medicine based on direct sources began in 1826, when agents working for antiquities dealer Giuseppe Passalacqua (1797–1865) unearthed numerous papyrus scrolls at Saqqara (Passalacqua 1826, p. 207). One of these, sold by Passalacqua in 1827 to Friedrich Wilhelm III of Prussia, is likely the papyrus known today as "Berlin 3038"; in other words, the first medical papyrus to be discovered and, later, translated. The importance of medical papyri as reference points for grasping the importance of Egyptian medicine and, above all, its role in the history of medical knowledge in general, however, only began a century later, with the publication of P. Edwin Smith, "a point of reference, showing for the first time an ancient Egyptian medical text almost free of magic and following rational lines of diagnosis and treatment" (Nunn 1996, p. 26).

Only a few medical papyri are known to have been accessed over the millennia, and they represent only a small fraction of the medical papyri that existed (David 2008a, p. 187). Each surviving papyrus reveals a particular aspect of reality in its own way, whether it be strictly medical, pathological or sanitary, and each one is striking in terms of the attention it pays to medical subjects.

These papyri testify as much to empirico-rational approaches as to magico-religious ideas on the origin, course and treatment of illnesses, and they "indicate a recurring three-pronged strategy for all healers, each with its own Egyptian term: *shesau*, treatment based on observation (diagnosis and prognosis), *pekheret*, medicines, made up of specific [...], and *ru* for formulas or incantations" (Quirke 2015, p. 194). In the papyri where remedies and recipes are listed, there is only the name of the disease or the part of the body affected, followed by the prescribed medicines and/or instructions for the preparation of the medicine and its application, generally expressed in short, passive verbal form.

Some appear highly structured, others less so, as if they were notes for personal use. The papyri are all written in hieratic and demotic script, with the exception of Ramesseum V (written in cursive hieroglyphs), and some of these seem to betray an older origin of almost a thousand years, as suggested by the presence of obsolete grammatical structures and glossaries explaining archaic terms. Numerous lists of medical papyri exist, and some even include medico-magical papyri[2]; we provide a list in Table 1.1 that includes dates and main bibliographical references.

2 See, for example, www.ucl.ac.uk/museums-static/digitalegypt/med/healingpapyri.html and www.medizinische-papyri.de/Start/html/medizinische_papyri.html (da Silva Veiga 2009, p. 10; Westendorf 1999, I, pp. 4–77).

Papyrus	Location	Date	First edition and main bibliography	Web link
Ramesseum V Papyrus	London, British Museum	Middle Kingdom, 12th Dynasty (around 1900 BCE)	(Gardiner 1955, pls. XV–XVII; Westendorf 1999, pp. 6–8)	www.webarchive.nationalarchives.gov.uk/20190801105848/https://www.britishmuseum.org/research/publications/online_research_catalogues/rp/the_ramesseum_papyri.aspx
Kahun Papyrus (UC 32057)	London, University College	Middle Kingdom, 12th Dynasty (around 1850 BCE)	(Griffith 1898, p. 5–11, pls. V–VI; Westendorf 1999, pp. 8–11)	www.ucl.ac.uk/museums-static/digitalegypt/med/birthpapyrus.html
Ramesseum III and IV Papyrus	London, British Museum	Middle Kingdom, 12th Dynasty (around 1800 BCE)	(Gardiner 1955, pls. VII-XIV; Westendorf 1999, pp. 11–15; Jean and Loyrette 2001, 2004)	www.webarchive.nationalarchives.gov.uk/20190801105848/https://www.britishmuseum.org/research/publications/online_research_catalogues/rp/the_ramesseum_papyri.aspx
Berlin 3027 Papyrus ("Zaubersprüche für Mutter und Kind")	Berlin, Ägyptisches Museum und Papyrussammlung	End of Hyksos period, beginning of New Kingdom (around 1550 BCE)	(Erman 1901; Westendorf 1999, p. 72; Yamazaki 2003)	–
Edwin Smith Papyrus	New York, The New York Academy of Medicine	Beginning of the New Kingdom (around 1550 BCE)	(Breasted 1930; Westendorf 1999, pp. 16–21; Allen 2005, pp. 70–115; Kosack 2011)	www.medizinische-papyri.de/PapyrusSmith/1280/index.html
Ebers Papyrus	Leipzig, Universitätsbibliothek	Beginning of the New Kingdom (around 1550 BCE)	(Ebers 1875; Westendorf 1999, pp. 22–35; Scholl 2002)	www.medizinische-papyri.de/PapyrusEbers/html/index.html

Papyrus	Location	Date	First edition and main bibliography	Web link
Hearst Papyrus	Berkeley, University of California - Bancroft Library	Early New Kingdom (around 1550 BCE)	(Reisner 1905; Westendorf 1999, pp. 35–37)	–
Louvre Papyrus E 4864 verso	Paris, Louvre	New Kingdom, 18th Dynasty (around 1430 BCE)	(Posener 1976, pp. 4–5, 144, pl. IV; Westendorf 1999, pp. 37–38)	www.medizinische-papyri.de/Start/html/-_papyrus_louvre_4864.html
Louvre Papyrus E 32 847	Paris, Louvre	New Kingdom, Amenhotep II (1424–1398 BCE)	(Bardinet 2018)	–
London Medical Papyrus (BM EA 10059)	London, British Museum	New Kingdom, 18th Dynasty (around 1350 BCE)	(Wreszinski 1912, pp. XIII-XIX 135–237; Westendorf 1999, pp. 38-41; Leitz 1999, pp. 51–84)	www.britishmuseum.org/collection/object/Y_EA10059-3
Brugsh Papyrus (P. Berlin 3038)	Berlin, Ägyptisches Museum und Papyrussammlung	New Kingdom, 19th Dynasty (around 1250 BCE)	(Brugsch 1863, pp. 101–120, pls. 85–107; Westendorf 1999, pp. 41–45)	www.medizinische-papyri.de/PapyrusBerlin3038/
Chester Beatty Papyrus VI (BM EA 10686)	London, British Museum	New Kingdom, 19th Dynasty (around 1250 BCE)	(Gardiner 1935, Vol. I, pp. 53–54 and Vol. II, pls. XXX-XXXII, XXXIA-XXXIIA; Westendorf 1999, pp. 45–48)	www.britishmuseum.org/collection/object/Y_EA10686-1

Papyrus	Location	Date	First edition and main bibliography	Web link
Papyrus Carlsberg VIII	University of Copenhagen, Egyptological Institute	New Kingdom, 19th–20th dynasties (around 1250–1150 BCE)	(Iversen 1939; Westendorf 1999, pp. 48–49)	–
Papyrus-Fragment Zagreb (inv. 881)	Zagreb, Archaeological Museum	New Kingdom, 19th–20th dynasties (around 1292–1070 BCE)	(Monnet Saleh 1970, p. 164; Westendorf 1999, p. 49; Uranić 2007, cat. 436)	www.medizinische-papyri.de/Start/html/-_papyrus_zagreb.html
Iatromagical Papyrus (P.Brooklyn 47.218.2)	New York, Brooklyn Museum	End of Pharaonic period/beginning of Ptolemaic period (5th–2nd century BCE)	(Guermeur 2016)	–
Rubensohn Papyrus (P.Berlin 10456)	Berlin, Ägyptisches Museum und Papyrussammlung	End of the Pharaonic period/beginning of Ptolemaic period (4th century BCE)	(Westendorf 1974; 1999, pp. 50–51)	www.medizinische-papyri.de/Start/html/-_papyrus_berlin_10456.html
Papyrus Brooklyn 47.218.48 and 47.218.85	New York, Brooklyn Museum	End of Pharaonic period/beginning of Ptolemaic period (4th century BCE)	(Sauneron 1989; Westendorf 1999, pp. 51–53; Aufrère 2012)	www.brooklynmuseum.org/opencollection/objects/60690 ; www.brooklynmuseum.org/opencollection/objects/124173
Berlin Demotic Papyrus from the Fayum (P.Berlin 13602)	Berlin, Ägyptisches Museum und Papyrussammlung	Ptolemaic period (end of 1st century BCE)	(Erichsen 1954; Westendorf 1999, pp. 53–54; Lieven and Quack 2018)	–

Papyrus	Location	Date	First edition and main bibliography	Web link
Demotic Herbal (P.Carslberg 230)	University of Copenhagen, Egyptian Institute	Roman period (2nd century CE)	(Tait 1991; Westendorf 1999, pp. 54–55)	–
Crocodilopolis Demotic Medical Book (P.Vindob. D.6257)	Wien, Österreichische *Nationalbibliothek*	Roman period (second half of the 2nd century CE)	(Reymond 1976; Westendorf 1999, pp. 55–58)	–
Demotic Papyrus-Fragment from Tebtynis	Oxford, Ashmolean Museum	Roman period (approx. 2nd century CE)	(Tait 1977, nr. 18, 61–65, pl. 5; Westendorf 1999, p. 58)	–
Demotic London Leiden Magical Medical Papyrus (BM EA 10 070 + P.Leiden I 383)	London, British Museum, Leiden, Rijksmuseum van Oudheden	Roman period (3rd century CE)	(Griffith and Thompson 1904–1909; Betz 1986; Westendorf 1999, pp. 75–76)	https://www.britishmuseum.org/collection/object/Y_EA10070-2
Wien Demotic Papyrus-Fragment (P.Vindob. D.12287)	Wien, Österreichische *Nationalbibliothek*	Roman period (3rd–4th centuries CE)	(Reymond 1984; Westendorf 1999, p. 59)	–

Table 1.1. *List of medical papyri*

Obviously, for each papyrus, there is a vast literature and, in many cases, a series of translations, as well as properly "medical" papyri ((Deines et al. 1958; Bardinet 1995) and online translation by Marko Stuhr[3]) and "magico-medical" papyri (among others (Leitz 1999; Beck 2018)). In addition to papyri, primary sources are also represented by short texts of cures and incantations written on ostraca dating from the 18th Dynasty to the time of the Roman conquest (Westendorf 1999, I, pp. 59–65).

There are also a number of surviving demotic papyri from this latter period, the most significant of which are the Crocodilopolis papyrus (P. Vindob 6257), rightly considered "an excellent testament to the adaptive nature of the Egyptian physician" (Ritner 2000, p. 113), and the London and Leiden magico-medical papyrus (P. London BM EA 10 070 + P. Leiden I 383). The latter, dated to the 3rd century CE, deserves special mention, as it is emblematic of the "syncretic and polymorphic culture" that emerges from its contents, as well as the linguistic mix of Egyptian, Greek and Coptic (Bresciani 1990, pp. 777–800). Coptic is used in many medico-magical texts as evidence of the extent to which Egyptian Christians had retained medical knowledge acquired in earlier periods (see, among others, Richter 2014; Bélanger Sarrazin 2017).

1.5. Diseases: a general assessment

As early as the beginning of the last century, studies by Marc Armand Ruffer (Ruffer 1921; Moodie 1923) revealed "that the Egyptians were constantly ill. The more he studied, the more illnesses he discovered" (Brier 2004, p. 19).

Today, advances in biomedical techniques and medical engineering have altered our ability to discover disease in ancient remains. Anatomical, radiological and paleopathological studies of mummified and skeletal remains – the main source of direct evidence of disease, and for this reason referred to as "biological archives" – provide a wealth of information to help us understand the pathology of these millennia: samples of tissue, bone, hair and teeth are examined using a range of techniques, such as CT scans, endoscopy, histology, immunocytochemistry, enzyme-linked immunosorbent assays and DNA analysis. We can say that science is now generally incorporated into studies of Egyptian history and archaeology and, combined with literary and artistic accounts, provides a complete picture of ancient diseases. The main ones were parasitosis, bacterial and viral infections, as well as tooth and eye diseases.

3 See: www.medizinische-papyri.de/Start/index.html.

Humans have always caught an incredible number of parasites, and "our knowledge of parasitic infections stretches back to antiquity, and descriptions of parasites and parasitic infections found in the earliest writings have been confirmed by the discovery of parasites in archaeological material" (Cox 2002, p. 595). Diseases caused by parasitic worms, in particular, were a major health problem for ancient Egyptians. Medical papyri contain unequivocal references to intestinal infestation by parasitic worms and, equally, to several identifiable remedies, even for specific worms (e.g. P. Ebers #50-85; P. Louvre E4864, 2, 4) of unquestionable deworming value, such as wormwood and pomegranate root bark (Nunn 1996, pp. 71–72). In ancient times, worms seemed to be generated by the corruption of organic matter, both postmortem and in vivo: the dual nature of this generatio aequivoca was corroborated by the observation of decomposing corpses, infected wounds and human feces in helminthiasis sufferers. Today, we know that most of these parasites were not only introduced through contaminated food, insect bites or tissue penetration – giving rise to ascariasis and taeniasis, and filariasis or strongyloidiasis, respectively – but also infested the waters of the Nile, giving rise to dracunculosis and schistosomiasis. The latter two parasitic diseases are worth noting, as "diseases carried by flies and parasites in rivers and irrigation canals represented a constant threat of blindness, disability and death" (Allen 2005, p. 9).

Now almost extinct, dracunculiasis (Guinea worm disease) was a widespread illness: a long, thin worm, *Dracunculus medinensis*, that could "appear" under the skin was more supernatural than natural, which contributed to the disease being reported since ancient times (in the Bible – Numbers, 21,6 – there is mention of the *ignitos serpentes* that afflicted the Israelites during the Exodus). Infection is caused by the ingestion of contaminated water containing copepods (*Cyclops sp.*, micro-crustaceans also known as water fleas) infected with parasites. A year after primary infection, the worm emerges as a cutaneous ulceration accompanied by laceration pain (generally on the ankle). In ancient times, the worm was extracted from the body by wrapping it around a stick. The operation was very complex and painful and could last up to several days, as it could provoke a severe anaphylactic reaction caused by toxins if the nematode broke loose. A possible reference to this delicate maneuver can be found in the P. Ebers (# 875) (Miller 1989).

Schistosomiasis "is successfully controlled in many countries, but remains a major public health problem" (Chitsulo et al. 2000; WHO 2002). The two most important trematodes (thin flatworms 1–3 cm long) responsible for this disease are *Schistosoma mansoni* (*S.m.*), for intestinal and hepatosplenic disease, and *Schistosoma haematobium* (*S.h.*). for genitourinary disease. Both have a life cycle in river water, and transmission of the disease occurs through prolonged immersion in fresh, non-flowing water contaminated by the feces and urine of parasitized individuals. In the case of chronic infestation, *S.m.* disease is characterized by abdominal pain and diarrhea, whereas in the case of *S.h.* disease, the most frequent symptom is hematuria. The Nile's land became a breeding ground for "the transmission and spread of schistosomiasis caused by *S.h.* after the development of irrigation farming in the early Pharaonic period" (Kloos and David 2002, p. 22). Some Egyptologists believe that the term used to describe it was *aAa* (Jonckheere 1944), a word that recurs in four medical papyri.

For both dracunculiasis and schistosomiasis, radiographic analysis reveals calcifications that are characteristic and pathognomonic of parasitic infestation in the abdominal wall and residual bladder tissue (Isherwood et al. 1979, p. 38). In 1910, paleopathology began with Ruffer's discovery of calcified *S.h.* eggs in the renal tubules of two Egyptian mummies from the 20th Dynasty (Ruffer 1910); today, scientific study continues using techniques such as immunocytochemistry and other molecular techniques, namely, enzyme-linked immunosorbent assays (ELISAs) and DNA analysis, to overcome the problems encountered when working with ancient samples (Rutherford 2016).

Stagnant water is a favorable condition for the formation of the pabulum where the Anopheles mosquito lived. In ancient Egypt, the presence of malaria, a parasitic disease caused by the protozoan *Plasmodium falciparum*, has been confirmed by paleo-immuno-diagnostic methods and molecular analyses (Nerlich et al. 2008). It has been present in Egypt since the Predynastic period (Rabino Massa et al. 2000). In particular, on the basis of recent clarification of the criteria for identifying malaria from bones applied to newly recovered material from Amarna, the gross prevalence rate for the northern cemetery sample is 72.4% (Dabbs and Rose 2015), and it cannot be ruled out that the early geological disruption and sudden population increase that followed the founding of Akhetaten may have altered the environment to the point of creating an optimal pabulum for the development of anophelines (Sabbahy 2021).

Caused by *Mycobacterium tuberculosis*, tuberculosis has always been widespread in ancient Egypt (in the form of *phtisis* – consumption), as the prolonged inhalation of silica dust, either from desert sand or from dust whirled up in stone quarries, favored the disease, in accordance with the fact that silicosis can increase the risk of phthisis. Silicosis and anthracosis (pneumoconiosis) can also be linked to oil lamp smoke. The disease has been attested since the early Dynastic period and is present on skeletal remains, with clear cases of ankylopoietic spondylitis, known as Pott disease (Bedeir 2004; Sabbahy 2014). Recent results of molecular analysis have confirmed that tuberculosis can be diagnosed in skeletal material dating back as far as 3000 BCE (Nerlich et al. 1997; Zink et al. 2001) and have demonstrated the presence of the disease, with infection rates of 20–30% in different populations of ancient Egypt, over a period of at least 2,500 years (Zink et al. 2003). In addition, molecular paleobiology suggests that tuberculosis may have been endemic, identifying the most likely time and place of its spread in the ancient Neolithic and Middle East, in what is known as the Fertile Crescent, extended to the Nile Valley, when human groups became more sedentary and numerous at that time and in that region, favoring its wide dissemination (Comas et al. 2013). In particular:

> Thirteen cases from the Nagada site in Upper Egypt (4500–3000 BCE) suggest that the first traces of tuberculosis in Egypt may date back to 4500 BCE. The first Egyptian cases of tuberculosis confirmed by molecular analysis date back to the Predynastic Period (3500–2650 BCE) and suggest that *M. tuberculosis* infection was relatively recurrent at that time (Buzic and Giuffra 2020).

Other bacterial or viral infections, such as leprosy (skeletal findings consistent with a diagnosis of leprosy are not confirmed beyond the Ptolemaic period; Dzierzykray-Rogalski 1980), tetanus, polio and smallpox, are very difficult to assess with certainty, as there are currently no reliable data to support paleopathological evidence. With respect to the plague, however, while only a critical assessment of supposed plague epidemics could be made until recently due to the inability to determine its exact nature, despite indications of bubonic disease in the middle of the 18th Dynasty (Habicht et al. 2021), researchers have recently found *Yersinia pestis* DNA in the bone tissue and intestinal contents of a mummified adult male from the Egyptian Museum in Turin. This mummy has been radiocarbon dated to the late Second Intermediate Period to the early New Kingdom (1686–1449 BCE) (Maixner et al. 2024). Another epidemic most likely broke out during the 14th year of Akhenaten's reign. Indeed, at Amarna, in addition to the evidence of paludism, the "high frequencies of subadults and young adults after weaning" in the cemetery and the existence of multiple burials suggest that Amarna "was plagued by an epidemic of massive proportions" (Dabbs and Zabecki 2014, pp. 236–38). It is no coincidence that, in P. Edwin Smith, we find eight spells "to ward off the breath of the annual plague" directed against the presumed carriers of a recurrent annual epidemic caused by morbid winds: an "evil air" quite different from the "gentle breath" of the north wind (P. Edwin Smith, 18.1–20.12).

Dental care was practiced because, apart from a low incidence of cavities until the first millennium BCE, Egyptians suffered from all of the dental pathologies resulting from the consumption of coarse foods: flours mixed with sand destroyed crowns through friction, leading to erosions and, above all, periodontal disease, a condition leading to tooth loss. According to paleopathological findings, "most of the teeth of the ancient Egyptian adults show extensive wear of the occlusal surfaces, and this wear, in a large number of cases, leads to the exposure of the dental pulp [...] when this results in the death of the pulp, bacteria invade the surrounding alveolar bone, which eventually becomes infected" (Leek 1967, pp. 51–52). Remedies for dental care are found in Ebers #739–749, in Ebers #553–555 and in prescriptions 8 and 9 of P. Hearst. However, the presence of dental surgery is dubious at best; in fact, "the papyri are silent on any operative aspects of dental care" (Nunn 1996, p. 205), and "there is no viable evidence to prove that there was any human intervention on tooth structure to relieve pain. The only active remedies that have been used have been applied to the surrounding soft tissue, and these remedies never delay the progression of dental disease" (Forshaw 2009).

Affected by light, irritated by sand and annoyed by insects, the Egyptians sought "preventive" protection in a black galena-based paste and a green malachite-based paste to shield their eyes from the Sun, sand and insects, which were sources of inflammation and infection. Painting the eyes "was not only a sacred rite, but part of a more or less rational treatment involving antibiotic elements" (Andersen 1997, p. 341); in particular, malachite powder is said to have inhibited bacterial growth (Majno 1975, pp. 111–115; Estes 1993, pp. 66–68). There are no paleopathological findings in the orbital region – apart, of course, from the relative frequency of porotic hyperostosis linked to severe anemia, ultimately caused by malnutrition and various kinds of infections, such as parasitic infections (Andersen 1997, p. 341) and especially malaria (Cerutti et al. 1999); however, the fact that eye diseases were very frequent is indirectly attested by the numerous prescriptions in the papyri and, in particular, by the "remedies for eye disorders" of P. Ebers, a compact and monothematic section ("Beginning of the medical compendium for the two eyes" #336, 55, 20) with 95 ad hoc prescriptions (Ebers #336–431). This part, although badly damaged, is copied without any variants in the slightly later P. Carlsberg VIII. In the texts, it is very difficult to identify specific diseases, whereas the symptoms described are precise and speak of dryness, bleeding, pain and inflammation. There are remedies for "eliminating blurring in the eye" (Ebers #339, 340, 415) and remedies for excessively watery eyes (Ebers #376, 378), passages that could perhaps attest to cataracts, not forgetting that "soon after Alexander introduced Egypt and Taxila (ancient India) to the Hellenic-speaking world, we find evidence that the Greeks became acquainted with cataract surgery" (Leffler et al. 2020). The most frequent and certainly the most serious infection was trachoma. This granulomatous conjunctivitis, caused by *Chlamydia trachomatis*, can lead to trichiasis (Ebers #424, 425) and blindness, although it can also result from other conditions. For the eyes, too, no mention is made of any type of surgical intervention, but a certain familiarity with wounds, bandaging and the removal of foreign bodies must have existed, as is also shown by the detail of a painting from the tomb of Ipuy (de Garis Davies 1927, pl. XXXVIII) at Deir el-Medina (TT 217, 20th Dynasty), as well as the texts (Ebers #337, 349, 381).

The most recent studies in the field of paleopathology have confirmed that malignant tumors have afflicted mankind since the earliest times (Giuffra et al. 2004, p. 81; Sabbahy 2015). As the frequency of cancerous diseases is generally linked more to life expectancy than to environmental or genetic factors (Nerlich et al. 2006), it is only natural that very little is known about their paleopathology and paleoepidemiology, and the history of cancer among ancient peoples continues to be the subject of study. While the upper social classes, benefiting from a more favorable lifestyle, could sometimes reach advanced ages, the average lifespan of Egyptian man was around 36 years (Masali and Chiarelli 1972; Toivari-Viitala 2001, p. 207). While evidence of primary malignant tumors is very rare in so-called biological archives (Capasso 2005, p. 6), it is less rare to find bone metastases, and in particular lytic metastases. Nevertheless, evidence of oncological diseases in medical papyri is still very uncertain, if not non-existent.

When examining the suffering man, "it was natural for the physician to assess the general condition of the patient from the quality of the pulse" (Sigerist 1951, p. 329), as shown by the passage with the so-called *Treatise on the Heart* begins (Ebers #854, analogous to Edwin Smith's Case 1, Gloss. A). The passage shows how the practitioner detects heart activity by examining the pulse at different points on the body and testifies to the way in which the relationship between the heart and the peripheral system was already understood (Nunn 1996, p. 113). Cardiac disorders were also studied, as witnessed, for example, by P. Ebers, which contains a series of glossaries explaining the terms used to describe pathological states of the heart, from the feeling of pressure to pericardial pain to collapse (Nunn 1996, p. 86, Table 4.2).

Paleopathology has long highlighted atheromatous vascular lesions in mummy tissues (Ruffer 1911), and the high frequency of atheromatous disease, both coronary and peripheral, has recently been demonstrated by X-ray (Finch 2011; Thompson et al. 2013). Given that atherosclerotic disease thus seems to have passed "freely" through four thousand years of human history, there are questions about the etiology of this disease in ancient Egypt and certainly not about the pathogenesis that is known today. In this particular environment, the triggering *vulnus* would have been "persistent infectious diseases", caused or transmitted by parasites (protozoa, as in the case of marsh fever, or helminths, as in the case of schistosomiasis and ascariasis), with an identical concatenation of pathogenic events ultimately leading to vascular lesions. If this were the case, where they were endemic, parasitic diseases would have been an unavoidable risk factor.

The brain was known to embalmers and physicians. Through their observation of head wounds, they also noticed that it was wrapped in a membrane (P. Edwin

Smith, Case 6, 2.24–25) with "an interesting use of the word to indicate a rupture of the meninges of the brain" (Breasted 1930, p. 157). The removal of the brain, followed by its brutal suppression by Egyptians during the embalming process, seems a clear indication that the importance of this organ is not understood, although some believe that "evidence from medical papyri and surviving mummies reveals that the Egyptians recognized the role of the brain" (Riggs 2010, p. 4).

As far as gynecology and obstetrics are concerned, it is worth noting that the Kahun papyrus, which contains only one incantation (No. 30) out of a total of 34 texts, is entirely devoted to women or, better still, that the "female matter" is enclosed in an ad hoc papyrus to such an extent that it would be natural to call it a "gynecological" papyrus. While the first seventeen cases are written with rigorous logic, without explicit magical references, in a rigid pattern of diagnosis and treatment, the remaining part deals with questions linked to pregnancy (in other words, fertility, conception, sex of the unborn child, tests to predict the course of pregnancy) and is thus evidence of the attempt to study, understand and manage it (Strouhal et al. 2010). Prognosis in the field of pregnancy is such a deeply felt need that even New Kingdom sources "contain an extraordinary series of texts relating to fertility, pregnancy and the determination of the sex of the unborn child" (Nunn 1996, p. 191), as well as P. Berlin 3038 #193–199 and the reverse of P. Carlsberg VIII. In P. Ebers, there are also "remedies that are made for women" (#783–837), with a large section also containing advice on childbirth (#797–807). In P. Ramesseum III and IV, we find passages on obstetrical and gynecological subjects in the broadest sense, with incantations to promote childbirth (Jean and Loyrette 2001, 2004). The fragment from papyrus Berlin 3027 is a compendium of apotropaic formulas for mother and child, defined as "the oldest treatise on pediatrics": although not strictly medical (it contains only three actual recipes as opposed to 14 pages of magic formulas), it is interesting because it refers to the care of mother and child "and looking inside the women's haven is generally more gratifying than learning, for the thousandth time, what can be cured with honey, beer and onions" (Erman 1901, p. 4).

Interest in the subject of obstetrics never ceased, as evidenced by P. Brooklyn 47.218.2 (5th–3rd centuries BCE), which contains a text of an obstetrical nature and primarily deals with pregnant woman and childbirth. The cautious, methodical pragmatism found in this papyrus is composed without opposition to magic, so much so that it can be said that "if in this text medicine and magic are mixed, it is in a very logical way: magic deals with protection, while prognosis and eventual treatment fall exclusively within the medical domain" (Guermeur 2013, p. 21). Here, too, no interventions are described, but indications are given for bone, topical and fumigation treatments. The recent discovery of a female mummy with clear signs of treatment in a tomb from the end of the 12th Dynasty in the Qubbet el-Hawa necropolis (QH34aa) (Jiménez-Serrano et al. 2020) confirms that the cures described were carried out. For paleopathological discoveries, let us recall the ovarian tumefaction of the Bozzi Granville mummy, which was first diagnosed as a malignant tumor and later as a benign cystadenoma (Donoghue 2010; Cosmacini and Cosmacini 2013, pp. 154–156), uterine prolapses (Nunn 1996, p. 196), the probable identification of a placenta (Mekota et al. 2005), and the Predynastic mummy of a young woman who died during childbirth at Turin's Museum of Anthropology and Ethnography (Chiarelli 1979, p. 58).

1.6. The field of trauma

In terms of frequency and importance, the numerous, mostly traumatic, injuries of the "workforce" requiring medical attention were certainly not little (Brandt-Rauf and Brandt-Rauf 1987). They were so frequent and so substantial that by the 2nd millennium BCE, traumatology was already taking the form of a methodical study: indeed, P. Edwin Smith reflects the experience that was largely gained on the battlefield and of injuries that could occur on building sites. The papyrus shows how "for the first time, the human mind peered into the mysteries of the human body and recognized conditions and processes as due to intelligible physical causes" (Breasted 1930, p. XIII): in practice, this is the precise description of 48 cases represented by male injuries consisting mainly of wounds, sprains, dislocations and fractures (injuries, in any case, always induced by trauma), following the *a capite ad calcem* (from top to bottom) exposition for the first time. In each case, we follow a clear, pragmatic, exhaustive and simple pattern, the synthesis of an analytical elaboration that took place "upstream". In their topographical layout, traumas are listed according to their increasing severity, and "symptoms were considered in groups (syndromes) rather than individually – a very important advance that is generally attributed to the Greeks" (David 2008a, p. 189).

While most other papyri can be said to correspond to each other over the centuries, in the case of P. Edwin Smith, no text is repeated in any other medical papyrus. Moreover, it contains 69 glossaries that form a dictionary of technical terms (Sigerist 1951, p. 308). It is also for these reasons that it has been called the *Book on Wounds*, *Book on Trauma*, *Surgical Treatise* or "the oldest scientific treatise in the world", not least because of its almost complete absence of a magico-religious aspect and its logical approach to the sufferer (Capart 1922).

P. Edwin Smith's surgery is what we call "minor surgery" today; in other words, maneuvers to reduce fractures and dislocations, such as those of the mandible (Case 25) and clavicle (Case 34). Among the most interesting cases is No. 8, where the author of the papyrus "noted the effect of the cranial injury on the lower limbs by referring to the side of the skull that received the injury" (Breasted 1930, p. 202). Whether this is a scribal error or an actual case of a secondary injury is still debated, but it is impossible not to note that "this case concludes with the observation that injuries to the brain and skull can affect the lower limbs, thus confirming an awareness of the localization of functions in the brain" (David 2018, p. 74).

Trauma was managed by the Egyptian physician, not only using specific maneuvers, but also by incision, suturing, dressing and bandaging; in particular, grease, honey and lint were the standard ointments for P. Edwin Smith's wounds (Majno 1975, p. 115; Bergman et al. 1983). It certainly seems that the therapeutic efficacy of honey as a wound-healing agent was known to Egyptians in two ways: on the one hand, through its application in cases of various traumas, where the risk of infection is increased (honey acts here as an antibiotic and antiseptic), and, on the other hand, as a follow-up therapy to keep the wound clean at all times. Similarly, in P. Edwin Smith's Cases 41 and 46, copper and sodium salts are used to reduce inflammation. A frequent topical remedy for wounds is fresh meat, which can promote coagulation (Estes 1993, p. 64).

Egyptians seem to have learned that a closed wound heals faster than an open wound: they had the technology to stick a dressing to the skin and invented the adhesive bandage (Forrest 1982, p. 200). Traumatological remedies are also found in P. Hearst and P. Ebers: in the former, there are indications for fractured limbs (#217–227) and wounds (#245–248); in the latter, there is an extensive section on remedies for burns and scalds (#482–509, with similarities in P. London EA10059), bruises and contusions (#510–514), as well as wounds (#515–#542).

The last part of P. Ebers is called the "Treatise on Tumors" (#857–877), as it lists 21 texts dealing with swellings and tumors in the broadest sense. For the latter, which are also the clinical expression of bacterial or viral infections, "knife" treatment is also expressly indicated as a complement to medicines.

Paleopathological findings include splints around fractured bones and examples of healed fractures with good alignment (Smith 1908; Ghalioungui 1963, Plate V; Zaki 2013). A much-discussed matter from the Temple of Horus the Elder and Sobek at Kom Ombo (dating from the end of the 2nd century CE) shows probable instrumentation (Porter and Moss 1939, p. 180, No. 228), and the aforementioned painting from the tomb of Ipuy at Deir el-Medina shows a series of occupational injuries and their treatment. Knives, blades and lancets, probes and styli, tweezers and needles are rare archaeological finds in ancient Egypt and would become more common in Greco-Roman Egypt (Marganne 1987), but the fact that the healer "desired a variety of choice in his instruments is evident in the medical papyri, which document three styles of bronze knives in addition to disposable – and therefore relatively sterile – flint or fine reed blades" (Ritner 2000, p. 114). Egyptians had different words for various knives, one of which was *hemem* cautery (Jonckheere 1951b). Some museums display an *instrumentarium* containing certain artifacts that were probably specially forged for surgical purposes (Jean 1999, 2012, 2015)[4]. The most recent discovery in this respect was made at Saqqara in the mastaba of Qar, "physician of the palace and keeper of the king's secrets", a physician who lived between the 5th and 6th dynasties" (Hawass 2004, pp. 202–203).

1.7. Healing remedies

A wide range of organic and inorganic materials were available to treat and hopefully cure (Deines and von Grapow 1959; Marganne and Koemoth 2003). Thanks to a wise, if empirical, confirmation of the efficacy of therapy, drugs, potions and spells were administered in every possible combination, and it is thus possible to see, even behind ritual activity, how "the evidence of a rich culture whose empirical knowledge is more extensive than might at first have been imagined, is concealed" (Rouffet 2013, p. 31). Moreover, given the relatively few effective medicines and surgical procedures available, it would be wrong to underestimate the healing value of the power of suggestion that would accompany the use of magic: it is wrong to consider suggestion irrelevant to the healing process, as clearly demonstrated today by the "placebo effect", which is recognized as having an assessable and sometimes measurable healing impact (Nunn 1996, p. 112; Cosmacini 2015, p. 56).

Unpleasant substances (such as excrement and urine) were thought to repel evil spirits, while aromatic substances were thought to attract benevolent spirits, which in turn could drive away evil ones (Estes 1993, p. 106). Thus, many materials acquired the merit of being life-giving and became "rational" tools for treating any life-threatening condition (Smith 1974, p. 19). The therapeutic act was directly conceived and managed by the complex figure of the healer, prefiguring the adage that "the medicine most often used is the physician himself" (Balint 1957, p. 1).

Egypt was the place "where the fertile earth produces powerful and diverse juices, some salubrious, others deadly" (*Odyssey*, Book IV, vv. 229–230), a fact confirmed by the wealth of materia medica in the medical literature. Among the

4 See, in particular, Paris (Louvre Museum and History of Medicine Museum), Cairo (Egyptian Museum), Saqqara (Imhotep Museum), Luxor (Mummification Museum) and Aswan (Nubian Museum).

large number of substances mentioned in medical papyri, some 160 plant products are recognized, of which only a fifth are accurately identified (Nunn 1996, p. 162; Ritner 2000, p. 113). Approximately 30 substances are of mineral origin (Nunn 1996, p. 146, Table 7.3), and there is a wide range of animal products. The identity of 55 plants and minerals is disputed, and 145 ingredients remain unidentified (Campbell 2008, p. 218). As a result, although the Egyptians possessed "potent juices", it is often difficult to understand what they intended to heal with.

The Egyptians exploited their natural environment to produce effective plant remedies, whose active compounds have recently begun to be isolated using modern techniques (McGovern et al. 2009, p. 7361). It must also be said that the innovative "scientific" aptitude of the Egyptian figurative tradition may have provided the discipline and method necessary for the development of phytography, whereas the presence of multiple sets of determinatives for plant names in the Egyptian graphic system may have formed the structure of a rudimentary taxonomy. Although we do not possess herbariums or systematic plant descriptions, we can assume that botanical reference texts existed as early as the middle of the 2nd century BCE (Betrò 2015). The plants of the "pharaonic flora" – which have been extensively studied since the last century (Loret 1892) – could be macerated, dried, cooked, crushed, mixed or filtered: a surprisingly high number of the plants used have medicinal properties recognized in modern phytotherapy, and even though the Elder healer did not fully understand the reason for their intrinsic usefulness, they certainly learned the best method for making their use more effective through experience. An initial attempt to estimate the pharmaceutical value of the substances used showed that in approximately two-thirds of cases, they were therapeutically effective and that while some treatments still relied on the power of magic, most followed rational "protocols" (David 2008b).

The various ingredients could be soaked in water, wine, vinegar, beer or oil to extract their active ingredients. Wine and other alcoholic beverages were an excellent way of dissolving herbal preparations, and wine itself contained enough alcohol to extract alkaloids such as morphine and codeine from poppies (although there is no archaeobotanical evidence of their use before the Roman period). Fruit trees were a source of sugar and thus of ethanol: the sweet liquid exuded by ripe fruit provided the ideal water and nutrient conditions for the yeasts on its surface to multiply and convert sugar into alcohol (McGovern et al. 2009, p. 7361). In addition to water, oil, fat, honey, milk, wine and beer served as carriers and binders; moreover, each was also considered a medicine in its own right (Metcalfe 2010).

The routes of administration were oral (tablets, infusions, decoctions, juices, syrups), ocular (drops), rectal (suppositories and enemas), vaginal, topical (sometimes with dressings) and by fumigation and inhalation (for the vagina and respiratory tract, respectively). The total dose is never expressed (perhaps because it is standardized), and each substance is indicated in proportion, sometimes according to the specific but simple use of fractions that refer to the different parts into which the Eye of Horus is divided.

The main medicine, secondary medicine and carrier or excipient substances are recognizable in 70% of prescriptions, whereas only 30% of remedies are characterized by polypharmacy (Campbell 2008, p. 232). In the latter cases, however, prescriptions may include so many substances in the same prescription that it is difficult to find, if at all, the medicine responsible for the positive therapeutic effect; as an example, see one of the prescriptions "to soften the *metw*", which contains 37 elements with no indication as to which is the most important (Ebers #663).

The use of the castor plant (*degem*) occupies a separate section of the P. Ebers (#251), where the use of its leaves, as well as its roots, seeds and oil, is proposed to treat headaches, abdominal disorders (it is a laxative), baldness and skin erythema. Castor oil was also one of the products that the artisans of Deir el-Medina received at regular intervals (Loret 1902, p. 698; Manniche 2006, pp. 150–151). Eight baskets of coriander seeds in Tutankhamun's tomb prove its importance in Egyptian culture and medicine: wine was fortified with herbs and, of the most likely herbal additives, only coriander is known by its ancient Egyptian name, *shaw* (McGovern et al. 2009, p. 7365). In P. Ebers, a mosquito repellent is recommended (Ebers No. 846). Two plants that could have been used in the Egyptian pharmacopoeia in various ways to prevent or treat malaria are *Moringa oleifera* (or *Moringa pterygosperma*, family Moringaceae) and *Artemisia annua* (family Asteraceae). *Artemisia absinthium* was used as an anthelmintic to expel roundworms and tapeworms. Traces of *Peganum harmala L.*, or wild rue (de Vartavan 1996, p. 11), which has not yet been identified in the Egyptian herbarium, have been found in Predynastic cultivation sites; endowed with psychoactive and oxytocic properties, it is identified by Dioscorides with *moly*, the prodigious plant mentioned by Homer in the *Odyssey* (X, 305), which was also used by the Copts (Manniche 2006, p. 153).

Already present at the dawn of Pharaonic civilization, organized beekeeping ensured the production of honey, which was – as mentioned – one of the fundamental medicinal substances. Healers were aware of its therapeutic efficacy and knew how and when to administer it. Honey is the ingredient most often mentioned in recipes (233 times in P. Ebers) and is the main remedy for coughs. Of the 40 cough remedies found in the papyri, 21 are found in P. Ebers (#305–325) under the title "Beginning of remedies to get rid of coughs" (#305, 53,1):

also included is a syrup made from figs, raisins and cumin (and more), to be mixed with beer, left overnight and drunk for four days (#321). Other recipes for treating coughs can be found in P. Hearst (#61) and P. Berlin 3038 (#29, 31–34, 36–47).

It is worth questioning whether it is still possible nowadays to benefit and learn from the traditional recipes of ancient Egypt (Pommerening 2019) in our era, at a time when, owing to the gradual disappearance of the languages used by indigenous peoples, we risk losing fundamental knowledge about the therapeutic properties of plants (Cámara-Leret and Bascompte 2021). This question remains open, and rediscovery within ancient texts is ongoing (Pommerening 2017). Furthermore, archaeobotany, phytogeography and pharmacognosy have made it possible to reassess prescriptions and their therapeutic efficacy, and analytical techniques, such as gas chromatography coupled with mass spectrometry (GC/MS) and liquid chromatography coupled with mass spectrometry (LC/MS), have added another aspect to the identification of the botanical and geographical origins of products (Campbell 2010).

The authors of Hippocratic treatises seem to have been particularly fond of Egyptian ingredients: for example, we find prescriptions for Egyptian white oil, Egyptian salt and Egyptian saffron, and other references to Egyptian medicines can easily be found in the Hippocratic corpus (Prioreschi 1996, p. 609). Texts containing references to Egyptian pharmacopoeia include the recipe collections used in Greco-Roman Egypt (Gazza 1956), and, according to Caelius Aurelianus, ancient Egyptian remedies were passed on to Dioscorides and Galen. After the Alexandrian era, the medicines first mentioned in Egyptian papyri continued to be used in the Western world, and "several ancient Egyptian medicines were still being used in the 19th century by physicians in Europe and America. Some, such as aloe and castor oil, survived into the 20th century" (Estes 1993, p. 113).

The Egyptians made an important contribution to traditional medicine through their pharmacological therapy and their skill in dealing with materia medica, so much so that the very symbol of drug prescription, namely, *Rx*, is said to derive from the Eye of Horus (Ritner 2000, p. 116). Anyway, to understand the vast and complex legacy of Egyptian pharmacopoeia to posterity, it could be sufficient to remember that the word φάρμακον most likely comes from *pekheret* (remedy) and *heka* (magic) (Aufrère and Marganne 2019), as if silently transposing an undeniably important part of Egyptian medical art into the act of healing of the present day physician.

1.8. References

Allen, J.P. (2005). *The Art of Medicine in Ancient Egypt*. Metropolitan Museum of Art/ Yale University Press, New York/New Haven/London.

Andersen, S.R. (1997). The eye and its diseases in ancient Egypt. *Acta Ophthalmologica Scandinavica*, 75(3), 338–344.

Anthes, R. (1928). *Die Felseninschriften von Hatnub nach den Aufnahmen Georg Möllers*. Hinrichs, Leipzig.

Assmann, J. (1990). *Ma'at: Gerechtigkeit und Unsterblichkeit im alten Ägypten*. C.H. Beck, München.

Assmann, J. (2005). *Death and Salvation in Ancient Egypt*. Cornell University Press, Ithaca.

Aufrère, S.H. (2012). Symptomatologie des morsures d'ophidiens d'après le papyrus Brooklyn nos 47.218.48 et 85 : aspects épistémologiques d'un texte égyptien ancien recopié au IVe siècle avant notre ère. *Anthropozoologica*, 47(1), 223–261.

Aufrère, S.H. and Marganne, M.-H. (2019). *Encounters Between Greek and Egyptian Science*. In *A Companion to Greco-Roman and Late Antique Egypt*, K. Vandorpe (ed.). John Wiley & Sons, Chichester, 501–518.

Austin, A. and Gobeil, C. (2017). Embodying the divine: A tattooed female mummy from Deir el-Medina. *Bulletin de l'Institut Français d'Archéologie Orientale*, 116, 23–46.

Balint, M. (1957). *The Doctor, His Patient and The Illness*. Pitman Medical Publishing, London.

Bardinet, T. (1995). *Les papyrus médicaux de l'Égypte pharaonique : traduction intégrale et commentaire*. Fayard, Paris.

Bardinet, T. (2018). *Médecins et magiciens à la cour du pharaon : une étude du papyrus médical Louvre E 32847*. Louvre/Khéops, Paris.

Beck, S. (2018). *Exorcism, Illness and Demons in an Ancient Near Eastern Context: The Egyptian Magical Papyrus I 343 + 345*. Sidestone, Leiden.

Bedeir, S.A. (2004). Tuberculosis in ancient Egypt. In *Tuberculosis*, Madkour, M.M. (ed.). Springer, Berlin, Heidelberg, 3–13.

Bélanger Sarrazin, R. (2017). Catalogue des textes magiques coptes. *Archiv für Papyrusforschung*, 63(2), 367–408.

Bergman, A., Yanai, J., Weiss, J., Bell, D., David, M.P. (1983). Acceleration of wound healing by topical application of honey. *American Journal of Surgery*, 145, 374–376.

Betrò, M. (2005). Religione. In *Egittologia*, Roccati, A. (ed.). Istituto poligrafico e Zecca dello Stato, Rome, 93–137.

Betrò, M. (2015). Gli erbari nell'antico Egitto. In *Naturalia e Artificialia I: "Le piante e i fiori d'Egitto nell'esperienza museografica degli scavi e degli erbari"*, Caporale, A., D'Amicone, E., Relats, F., Tragan, P.-R. (eds). Publications de l'Abadia de Montserrat, Barcelona, 41–60.

Betz, H.D. (1986). *The Greek Magical Papyri in Translation, Including the Demotic Spells.* The University of Chicago Press, Chicago/London.

Brandt-Rauf, P.W. and Brandt-Rauf, S.I. (1987). History of occupational medicine: Relevance of Imhotep and the Edwin Smith papyrus. *British Journal of Industrial Medicine*, 44(1), 68–70.

Breasted, J.H. (1930). *The Edwin Smith Surgical Papyrus Published in Facsimile and Hieroglyphic Transliteration with Translation and Commentary*. The University of Chicago Press, Chicago.

Bresciani, E. (1990). *Letteratura e poesia dell'antico Egitto*. Einaudi, Turin.

Brier, R. (2004). Infectious diseases in ancient Egypt. *Infectious Disease Clinics of North America*, 18(1), 17–27.

Brugsch, H. (1853). Über die medicinischen Kenntnisse der alten Ägypter und über ein alt-ägyptisches medicinisches Manuscript im Königl. Museum zu Berlin. *Allgemeine Monatsschrift für Wissenschaft und Literatur*, 44–56.

Butzer, K.W. (1976). *Early Hydraulic Civilization in Egypt: A Study in Cultural Ecology. Prehistoric Archeology and Ecology*. The University of Chicago Press, Chicago/London.

Buzic, I. and Giuffra, V. (2020). The paleopathological evidence on the origins of human tuberculosis: A review. *Journal of Preventive Medicine and Hygiene*, 61(1), E3–E8.

Cámara-Leret, R. and Bascompte, J. (2021). Language extinction triggers the loss of unique medicinal knowledge. *PNAS*, 118(24), e2103683118.

Campbell, J.M. (2008). Pharmacy in ancient Egypt. In *Egyptian Mummies and Modern Science*, David, R. (ed.). Cambridge University Press, Cambridge/New York, 216–233.

Campbell, J.M. and David, A.R. (2010). The application of archaeobotany, phytogeography and pharmacognosy to confirm the pharmacopoeia of Ancient Egypt 1850–1200 B.C. In *Pharmacy and Medicine in Ancient Egypt: Proceedings of the Conferences held in Cairo (2007) and Manchester (2008)*, Cockitt, J. and David, R. (eds). Archaeopress, Oxford, 20–29.

Capart, J. (1922). Le plus vieux traité scientifique du monde. *Bulletin de la Classe des lettres et des sciences morales et politiques*, 5(8), 338–340.

Capasso, L. (2005). Antiquity of cancer. *International Journal of Cancer*, 113(1), 2–13.

Cerutti, N., Marin, A., Rabino Massa, E., Savoia, D. (1999). Immunological investigation of malaria and new perspectives in paleopathological studies. *Bollettino della Società Italiana di Biologia Sperimentale*, 75(3/4), 17–20.

Chiarelli, B. (1979). Paleobiologia degli egizi. *Le Scienze*, 132, 56–64.

Chitsulo, L., Engels, D., Montresor, A., Savioli, L. (2000). The global status of schistosomiasis and its control. *Acta Tropica*, 77(1), 41–51.

Cole, C. (1987). The role of women in the medical practice of ancient Egypt. *Discussion in Egyptology*, 9, 25–29.

Collier, M. and Quirke, S. (2004). *The UCL Lahun Papyri: Religious, Literary, Legal, Mathematical and Medical*. Archaeopress, Oxford.

Comas, I., Coscolla, M., Luo, T., Borrell, S. Holt, K.E., Kato-Maeda, M., Parkhill, J., Malla, B., Berg, S., Thwaites, G. et al. (2013). Out-of-Africa migration and Neolithic coexpansion of *Mycobacterium tuberculosis* with modern humans. *Nature Genetics*, 45, 1176–1182.

Contin, F. (2005). La medicina nell'antico Egitto. *Antrocom*, 1(2), 115–120.

Cosmacini, P. (2015). *Il medico d'oggi è nato in Egitto. Alle origini del pensiero medico moderno*. Piccin, Padua.

Cosmacini, G. and Cosmacini, P. (2013). *Il medico delle mummie. Vita e avventure di Augustus Bozzi Granville*. Laterza, Bari-Roma.

Cox, F.E.G. (2002). History of human parasitology. *Clinical Microbiology Reviews*, 15(4), 595–612.

Dabbs, G.R. and Rose, J.C. (2015). The presence of malaria among the non-royal of the North Tombs Cemetery. *Horizon*, 16, 7.

Dabbs, G.R. and Zabecki, M. (2014). Abandoned memories: A cemetery of forgotten souls? In *Remembering the Dead in the Ancient Near East*, Porter, B.W. and Boutin, A.T. (eds). University Press of Colorado, Boulder, CO, 217–250.

David, R. (2004). Rationality versus irrationality in Egyptian medicine in the Pharaonic and Graeco-Roman Periods. In *Magic and Rationality in Ancient Near Eastern and Graeco-Roman Medicine*, Horstmanshoff, H.F.J. and Stol, M. (eds). Brill, Leiden/Boston, 133–151.

David, R. (2008a). The ancient Egyptian medical system. In *Egyptian Mummies and Modern Science*, David, R. (ed.). University Press, Cambridge, 181–194.

David, R. (2008b). The art of healing in ancient Egypt: A scientific reappraisal. *Lancet*, 372(9652), 1802–1803.

David, R. (2018). Egyptian medicine. In *The Oxford Handbook of Science and Medicine in the Classical World*, Keyser, P.T. and Scarborough, J. (eds). Oxford University Press, Oxford, 71–79.

Dawson, W.R. (1942). Medicine. In *The Legacy of Egypt*, Glanville, S.R.K. (ed.). Clarendon, Oxford, 179–197.

Deines, H. and von Grapow, H. (1959). *Grundriss der Medizin der alten Ägypter*, volume VI. Akademie-Verlag, Berlin.

Deines, H., von Grapow, H., Westendorf, W. (1958). *Grundriß der Medizin der alten Ägypter*, volume IV. Akademie-Verlag, Berlin.

Dolzani, C. (1993). I testi medici egiziani tra scomparsa e riscoperta. Possibili vie di un iter sotterraneo. In *VI Congresso Internazionale di Egittologia: Atti*. Tipografia Torinese Stabilimento Poligrafico, Turin, 107–113.

Donoghue, H.D., Lee, O.Y.-C., Minnikin, D.E., Besra, G.S., Taylor, J.H., Spigelman, M. (2010). Tuberculosis in Dr Granville's Mummy: A molecular re-examination of the earliest known Egyptian mummy to be scientifically examined and given a medical diagnosis. *Proceedings of the Royal Society*, 277(1678), 51–56.

Duquesne, T. (2002). La deification des parties du corps : correspondances magiques et identification avec les dieux dans l'Égypte ancienne. In *La magie en Égypte : À la recherche d'une definition : actes du colloque organisé par le Musée du Louvre les 29 et 30 septembre 2000*, Koenig, Y. (ed.). La Documentation française, Paris, 237–271.

Dzierzykray-Rogalski, T. (1980). Paleopathology of the Ptolemaic inhabitants of Dakhleh Oasis (Egypt). *Journal of Human Evolution*, 9(1), 71–72.

Ebers, G. (1875). *Papyrus Ebers: das hermetische Buch über die Arzeneimittel der alten Ägypter in hieratischer Schrift*. Wilhelm Engelmann, Leipzig.

Ejsmond, W., Ożarek-Szilke, M., Jaworski, M., Szilke, S. (2021). A pregnant ancient Egyptian mummy from the 1st century BC. *Journal of Archaeological Science*, 132, 105371.

Erichsen, W. (1954). Aus einem demotischen Papyrus über Frauenkrankheiten. *Mitteilungen des Instituts für Orientforschung*, 2, 363–377.

Erman, A. (1901). *Zaubersprüche für Mutter und Kind: aus dem Papyrus 3027 des Berliner Museums*. Verlag der Königlichen Akademie der Wissenschaften, Berlin.

Erman, A. and Grapow, H. (1940–1963). *Wörterbuch der ägyptischen Sprache*, 12 volumes. Akademie-Verlag, Leipzig/Berlin.

Estes, J.W. (1993). *The Medical Skills of Ancient Egypt,* revised edition. Science History Publications, Canton.

Étienne, M. (2016). La médicine et son exercice. In *Savoir et pouvoir à l'époque de Ramsès II: Khâemouaset, le prince archéologue : savoir et pouvoir à l'époque de Ramsès II*, Charron, A. and Barbotin, C. (eds). Musée départemental Arles antique/Snoeck, Arles/Gand, 255–257.

Eyre, C. (2013). *The Use of Documents in Pharaonic Egypt*. Oxford University Press, Oxford.

Filer, J. (1996). *Disease*. University of Texas Press, Austin, TX.

Finch, C.E. (2011). Atherosclerosis is an old disease: Summary of the Ruffer Centenary Symposium. The paleocardiology of Ancient Egypt, a meeting report of the Horus Study team. *Experimental Gerontology*, 46(11), 843–846.

Forrest, R.D. (1982). Early history of wound treatment. *Journal of the Royal Society of Medicine*, 75, 198–205.

Forshaw, R.J. (2009). The practice of dentistry in ancient Egypt. *British Dental Journal*, 206(9), 481–486.

Forshaw, R.J. (2014a). Before Hippocrates: Healing practices in ancient Egypt. In *Medicine, Healing and Performance*, Gemi-Iordanou, E., Gordon, S., Matthew, R., McInnes, E., Pettitt, R. (eds). Oxbow Books, Oxford, 25–41.

Forshaw, R.J. (2014b). *The Role of the Lector (Xry-Hbt) in Ancient Egyptian Society*. Archaeopress, Oxford.

Gardiner, A.H. (1935). *Hieratic papyri in the British Museum: Third Series: Chester Beatty Gift*. British Museum, London.

Gardiner, A.H. (1955). *The Ramesseum papyri*. Oxford University Press for Griffith Institute, Oxford.

de Garis Davies, N. (1927). *Two Ramesside Tombs at Thebes*. The Metropolitan Museum of Art, New York.

Gazza, V. (1956). Prescrizioni mediche nei papiri dell'Egitto greco-romano, II. *Aegyptus*, 36(1), 73–114.

Ghalioungui, P. (1963). *Magic and Medical Science in Ancient Egypt*. Hodder and Stoughton, London.

Ghalioungui, P. (1975). Les plus anciennes femmes-médecins de l'histoire. *Bulletin de l'Institut Français d'Archéologie Orientale*, 75, 159–164.

Giuffra, V., Ciranni, R., Fornaciari, G. (2004). I tumori maligni nell'Antico Egitto e in Nubia. *Egitto e Vicino Oriente*, 27, 81–93.

Gordon, A.H. and Schwabe, C.W. (2004). *The Quick and the Dead: Biomedical Theory in Ancient Egypt*. Brill/Styx, Leiden/Boston.

Griffith, F.L.I (1898). *The Petrie papyri: Hieratic papyri from Kahun and Gurob (Principally of the Middle Kingdom)*. Bernard Quaritch, London.

Griffith, F.L.I. and Thompson, H. (1904–1909). *The Demotic Magical Papyrus of London and Leiden*. Grevel, London.

Guermeur, I. (2013). Entre magie et médecine : l'exemple du Papyrus Brooklyn 47.218.2. *Égypte, Afrique & Orient*, 71, 11–22.

Guermeur, I. (2016). Le papyrus hiératique iatromagique n° 47.218.2 du Musée de Brooklyn. *Bulletin de la Société Française d'Égyptologie*, 193–194, 10–28.

Habicht, M.E., Eppenberger, P.E., Rühli, F. (2021). A critical assessment of proposed outbreaks of plague and other epidemic diseases in Ancient Egypt. *International Journal of Infectious Diseases*, 103, 217–219.

Hawass, Z. (2004). *Hidden Treasures of Ancient Egypt: Unearthing the Masterpieces of Egyptian History*. National Geographic, Washington, D.C.

Isherwood, I., Jarvis, H., Fawcitt, R.A. (1979). Radiology of the Manchester mummies. In *Manchester Museum Mummy Project: Multidisciplinary Research on Ancient Egyptian Mummified Remains*, David, A.R. (ed.). Manchester Museum, Manchester, 25–64.

Iversen, E. (1939). *Papyrus Carlsberg no. VIII: With Some Remarks on the Egyptian Origin of Some Popular Birth Prognoses*. Ejnar Munksgaard, Copenhagen.

Jean, R.-A. (1999). *À propos des objets égyptiens conservés au Musée d'Histoire de la Médecine*. Université René Descartes, Paris.

Jean, R.-A. (2012). *La chirurgie en Égypte ancienne; à propos des instruments médicochirurgicaux métalliques égyptiens conservés au Musée du Louvre*. Cybèle, Paris.

Jean, R.-A. (2015). Trois objets égyptiens antiques du Musée du Val-de-Grâce. *Clystère*, 38, 3–14.

Jean, R.-A. and Loyrette, A.-M (2001). À propos des textes médicaux des papyrus du Ramesseum Nos III et IV – I : la contraception. *Memnonia*, 12/13, 83–115.

Jean, R.-A. and Loyrette, A.-M. (2004). À propos des textes médicaux des papyrus du Ramesseum Nos III et IV – II : la gynécologie. *Memnonia*, 15, 67–91.

Jiménez-Serrano, A., Martínez-Martos, R., Barba-Colmenero, V., Alba-Gómez, J.M. (2020). A gynecological treatment found in Qubbet el-Hawa and described in different papyri. *Zeitschrift für ägyptische Sprache und Altertumskunde*, 147(2), 172–179.

Jonckheere, F. (1944). *Une maladie égyptienne ; l'hématurie parasitaire*. Fondation Egyptologique Reine Elisabeth, Brussels.

Jonckheere, F. (1951a). À la recherche du chirurgien égyptien. *Chronique d'Égypte*, 26(51), 28–45.

Jonckheere, F. (1951b). Le cautère est une invention pharaonique. *La vie Médicale*, 7, 3–5.

von Känel, F. (1984). *Les prêtres-ouâb de Sekhmet et les conjurateurs de Serket*. Presses Universitaires de France, Paris.

Kloos, H. and David, R. (2002). The Paleoepidemiology of Schistosomiasis in Ancient Egypt. *Human Ecology Review*, 9(1), 14–25.

Kosack, W. (2011). *Der medizinische Papyrus Edwin Smith: the New York Academy of Medicine, Inv. 217. Neu in Hieroglyphen übertragen, übersetzt und bearbeitet*. C. Brunner, Berlin.

Leek, F.F. (1967). The practice of dentistry in ancient Egypt. *The Journal of Egyptian Archaeology*, 53, 51–58.

Lefebvre, G. (1956). *Essai sur la médecine égyptienne de l'époque pharaonique*. Presses Universitaires de France, Paris.

Leffler, C.T., Klebanov, A., Samara, W.A., Grzybowski, A. (2020). The history of cataract surgery: From couching to phacoemulsification. *Annals of Translational Medicine*, 8(22), 1551. doi: 10.21037/atm-2019-rcs-04.

Leitz, C. (1999). *Magical and Medical Papyri of the New Kingdom*. The British Museum Press, London.

von Lieven, A. and Quack, J.F. (2018). Ist Liebe eine Frauenkrankheit? Papyrus Berlin P 13602, ein gynäkomagisches Handbuch. In *Hieratic, Demotic and Greek Studies and Text Editions: Of Making Many Books There is No End. Festschrift in Honour of Sven P. Vleeming*, Donker van Heel, K., Hoogendijk, F.A.J., Martin, C.J. (eds). Brill, Leiden/Boston, 257–274.

Lopez, F. (2015). *Democede di Crotone e Udjahorresnet di Saïs*. Pisa University Press, Pisa.

Lord, C. (2011). The veterinary papyrus of Kahun. In *Current Research in Egyptology 2009: Proceedings of the Tenth Annual Symposium, University of Liverpool 2009*, Corbelli, J., Boatright, D., Malleson, C. (eds). Oxbow, Oxford, 99–105.

Lord, C. (2016). One and the same? An investigation into the connection between veterinary and medical practice in ancient Egypt. In *Mummies, Magic and Medicine in Ancient Egypt: Multidisciplinary Essays for Rosalie David*, Price, C., Forshaw, R., Chamberlain, A., Nicholson, P., Morkot, R., Tyldesley, J. (eds). Manchester University, Manchester, 142–156.

Loret, V. (1892). *La flore pharaonique d'après les documents hiéroglyphiques et les spécimens découverts dans les tombes*. Leroux, Paris.

Loret, V. (1902). Le ricin et ses emplois médicinaux dans l'Ancienne Egypte. *Revue de Médecine*, 22(8), 687–698.

Lucarelli, R. (2017). Illness as divine punishment: The nature and function of the disease-carrier demons in the ancient Egyptian magical texts. In *Demons and Illness from Antiquity to the Early-Modern Period*, Bhayro, S. and Rider, C. (eds). Brill, Leiden, 53–60.

Lucarelli, R. (2020). The magician as a literary figure in ancient Egyptian texts. In *Cult Practices in Ancient Literatures: Egyptian, Near Eastern and Graeco-Roman Narratives in a Cross-Cultural Perspective. Proceedings of a Workshop at the Institute for the Study of the Ancient World, New York*, Naether, F. (ed.). ISAW Paper 18 [Online]. Available at: http://hdl.handle.net/2333.1/s4mw6x82.

Maixner, F., Gaul, E., Mussauer, A., Paladin, A., Wurst, C., Sarhan, M., Samadelli, M., Del Vesco, P., Facchetti, F., Posth, C. et al. (2024). The 3,300-year-old Yersinia pestis genome of an ancient Egyptian mummified individual. In *European Meeting of the Paleopathology Association PPA (Leiden, 21.8.2024 – 24.8.2024).*

Majno, G. (1975). *The Healing Hand. Man and Wound in the Ancient World.* Harvard University Press, Cambridge/London.

Manniche, L. (2006). *An Ancient Egyptian Herbal.* The American University in Cairo Press, Cairo.

Marganne, M.-H. (1987). *Les instruments chirurgicaux de l'Égypte grécoromaine*. Éditions APDCA, Juan-les-Pins.

Marganne, M.-H. and Koemoth, P. (2003). Pharmacopoea Aegyptia et Graeco-Aegyptia. CeDoPaL. Report [Online]. Available at: http://web.philo.ulg.ac.be/cedopal/pharmacopoea-aegyptia-et-graeco-aegyptia-anglais/.

Masali, M. and Chiarelli, B. (1972). Demographic data on the remains of ancient Egyptians. *Journal of Human Evolution*, 1(2), 161–169.

McGovern, P.E., Mirzolan, A., Hall, G.R. (2009). Ancient Egyptian herbal wines. *Proceedings of the National Academy of Sciences of the United States of America*, 106(18), 7361–7366.

Mekota, A.-M., Grupe, G., Zimmerman, M.R., Vermehren, M. (2005). First identification of an ancient Egyptian mummified human placenta. *International Journal of Osteoarchaeology*, 15(1), 51–60.

Metcalfe, R.J. (2010). Good for what ales you: A prospective study into the role of beer in ancient Egyptian medicine. In *Pharmacy and Medicine in Ancient Egypt, Proceedings of the Conferences held in Cairo (2007) and Manchester (2008)*, Cockitt, J. and David, R. (eds). Archaeopress, Oxford, 108–109.

Miller, R.L. (1989). Dqr, spinning and treatment of Guinea worm in P. Ebers 875. *Journal of Egyptian Archaeology*, 75, 249–254.

Monnet Saleh, J. (1970). *Les antiquités égyptiennes de Zagreb : catalogue raisonné des antiquités égyptiennes conservées au Musée Archéologique de Zagreb en Yougoslavie*. Mouton, Paris.

Moodie, R.L. (1923). *Paleopathology. An Introduction to the Study of Ancient Evidences of Disease*. University of Illinois Press, Urbana.

Moscati, S. (1975). *Apparenza e realtà nell'arte del Vicino Oriente antico*. Accademia Nazionale dei Lincei, Rome.

Nerlich, A.G., Haas, C.J., Zink, A., Szeimies, U., Hagedorn, H.G. (1997). Molecular evidence for tuberculosis in an ancient Egyptian mummy. *Lancet*, 350(9088), 1404.

Nerlich, A.G., Rohrbach, H., Bachmeier, B., Zink, A. (2006). Malignant tumors in two ancient populations: An approach to historical tumor epidemiology. *Oncology Reports*, 16(1), 197–202.

Nerlich, A.G., Schraut, B., Dittrich, S., Jelinek, T., Zink. A.R. (2008). Plasmodium falciparum in Ancient Egypt. *Emerging Infectious Diseases*, 14(8), 1317–1319.

Nunn, J.F. (1996). *Ancient Egyptian Medicine*. British Museum Press, London.

Nyord, R. (2009). *Breathing Flesh: Conceptions of the Body in the Ancient Egyptian Coffin Texts*. Museum Tusculanum Press, Copenhagen.

Passalacqua, J. (1826). *Catalogue raisonné et historique des antiquités découvertes en Égypte*. Imprimerie de C.J. Trouve, Paris.

Pommerening, T. (2017). Classification in Ancient Egyptian medical formulae and its role in re-discovering comprehensive and specific concepts of drugs and effects. In *Classification from Antiquity to Modern Times: Sources, Methods, and Theories from an Interdisciplinary Perspective*, Pommerening, T. and Bisang, W. (eds). De Gruyter, Berlin/Boston, 167–195.

Pommerening, T. (2019). Von der Textquelle zur Interpretation: Können wir heute noch von überlieferten Rezepturen aus altägyptischer Zeit profitieren? *Geschichte der Pharmazie*, 71, 11–20.

Porter, B. and Moss, R.L.B. (1939). *Topographical Bibliography of Ancient Egyptian Hieroglyphic Texts, Reliefs and Paintings VI. Upper Egypt: Chief Temples*, Griffith Institute, Oxford.

Posener, G. (1976). *L'enseignement loyaliste : sagesse égyptienne du Moyen Empire*. Librairie Droz, Geneva.

Prioreschi, P. (1996). *A History of Medicine: Greek Medicine, Volume II*, 2nd edition. Horatius Press, Omaha.

Quirke, S. (2015). *Exploring Religion in Ancient Egypt*. Wiley Blackwell, Chichester/Malden.

Rabino Massa, E., Cerutti, N., Marin, D., Savoia, A. (2000). Malaria in ancient Egypt: Paleoimmunological investigation on predynastic mummified remains. *Chungará (Arica)*, 32(1), 7–9.

Reisner, G.A. (1905). *The Hearst Medical Papyrus: Hieratic Text in 17 Facsimile Plates in Collotype with Introduction and Vocabulary*. Hinrichs, Leipzig.

Reymond, E.A.E. (1976). *From the Contents of the Libraries of the Suchos Temples in the Fayyum I: A Medical Book from Crocodilopolis: P. Vindob. D. 6257*. Brüder Hollinek, Vienna.

Reymond, E.A.E. (1984). From an ancient Egyptian dentist's handbook: P. Vindob. D. 12287. In *Mélanges Adolphe Gutbub*. Université de Montpellier, Montpellier, 183–201.

Richter, T.S. (2014). Neue koptische medizinische Rezepte. *Zeitschrift für ägyptische Sprache und Altertumskunde*, 141, 154–194.

Riggs, C. (2010). Body. In *UCLA Encyclopedia of Egyptology*, Frood, E. and Wendrich, W. (eds). Department of Near Eastern Languages and Cultures UCLA, Los Angeles, 3–15 [Online]. Available at: http://escholarship.org/uc/item/8f21r7sj.

Ritner, R.K. (1992). Implicit models of cross-cultural interactions, a question of noses, soap and prejudice. In *Life in a Multi-Cultural Society: Egypt from Cambyses to Constantine and Beyond*, Johnson, J.H. (ed.). Oriental Institute of the University of Chicago, Chicago, 283–290.

Ritner, R.K. (2000). Innovations and adaptations in Ancient Egyptian Medicine. *Journal of Near Eastern Studies*, 59, 107–117.

Ritner, R.K. (2001). Magic in medicine. In *The Oxford Encyclopedia of Ancient Egypt*, volume II, Redford, D.B. (ed.). Oxford University Press, Oxford, 326–329.

Robinson, J-M. (2017). 'Your mouth speaks, your feet walk': Representations of physical abnormalities and perceptions of the human body in ancient Egyptian funerary iconography. *Archaeological Review from Cambridge*, 32(1), 6–33.

Rouffet, F. (2013). Les ingrédients dans les prescripions medico-magiques égyptiennes. *Égypte Afrique & Orient*, 71, 23–32.

Ruffer, M.A. (1910). Note on the presence of "Bilharzia Haematobia" in Egyptian mummies of the twentieth dynasty (1250–1000 B.C.). *British Medical Journal*, 1, 16.

Ruffer, M.A. (1911). On arterial lesions found in Egyptian mummies (1580 BC–535 AD). *The Journal of Pathology and Bacteriology*, 16, 453–462.

Ruffer, M.A. (1921). *Studies in the Paleopathology of Egypt*. University of Chicago Press, Chicago.

Rutherford, P. (2016). Schistosomiasis, ancient and modern: The application of scientific techniques to diagnose the disease. In *Mummies, Magic and Medicine in Ancient Egypt: Multidisciplinary Essays for Rosalie David*, Price, C., Forshaw, R., Chamberlain, A., Nicholson, P.T. (eds). Manchester University, Manchester, 197–209.

Sabbahy, L. (2014). An overview of the evidence for tuberculosis from Ancient Egypt. In *Paleopathology in Egypt and Nubia: A Century in Review*, Metcalfe, R., Cockitt, J., David, R. (eds). Archaeopress, Oxford, 51–55.

Sabbahy, L. (2015). A decade of advances in the paleopathology of the ancient Egyptians. In *Egyptian Bioarchaeology: Humans, Animals, and Environment*, Ikram, S., Kaiser, J., Walker, R. (eds). Sidestone Press, Leiden, 113–117.

Sabbahy, L. (2021). Did Akhenaten's founding of Akhetaten cause a malaria epidemic? *Journal of the American Research Center in Egypt*, 56, 175–179.

Sauneron, S. (1989). *Un traité égyptien d'ophiologie : Papyrus du Brooklyn Museum No. 47.218.48 et .85*. Institut Français d'Archéologie Orientale, Cairo.

Scholl, R. (2002). *Der Papyrus Ebers, die größte Buchrolle zur Heilkunde Altägyptens*. Universitätsbibliothek, Leipzig.

Schwabe, C.W. (1986). Bull semen and muscle ATP: Some evidence of the dawn of medical science in ancient Egypt. *Canadian Journal of Veterinary Research*, 50, 145–153.

Sigerist, H.E. (1951). *A History of Medicine. Vol. 1: Primitive and Archaic Medicine*. Oxford University Press, New York.

da Silva Veiga, P.A. (2009). *Health and Medicine in Ancient Egypt: Magic and Science*. Archaeopress, Oxford.

Smith, G.E. (1908). The most ancient splints. *British Medical Journal*, 1, 732–736.

Smith, G.E. (1974). Introduction. In *Ancient Egyptian Medicine. The Papyrus Ebers*, Bryan, C.P. (ed.). Ares, Chicago.

Strouhal, E., Vachala, B., Vymazalová, H. (2010). *The Medicine of the Ancient Egyptians. 1: Surgery, Gynecology, Obstetrics and Paediatrics*. The American University in Cairo Press, Cairo/New York.

Tait, W.J. (1977). *Papyri from Tebtunis in Egyptian and Greek (P. Tebt. Tait)*. Egypt Exploration Society, London.

Tait, W.J. (1991). P. Carlsberg 230: Eleven fragments from a demotic herbal. In *The Carlsberg Papyri 1: Demotic Texts from the Collection*, Frandsen, P.J. (ed.). University of Copenhagen/Museum Tusculanum Press, Copenhagen, 47–92.

Thompson, R.C., Allam, A.H., Lombardi, G.P., Wann, L.S., Sutherland, M.L., Sutherland, J.D., Al-Tohamy Soliman, M., Frohlich, B., Mininberg, D.T., Monge, J.M. et al. (2013). Atherosclerosis across 4000 years of human history: The Horus study of four ancient populations. *Lancet*, 381(9873), 1211–1222.

Toivari-Viitala, J. (2001). *Women at Deir el-Medîna. A Study of the Status and Roles of the Female Inhabitants in the Workmen's Community during the Ramesside Period.* Nederlands Instituut voor het Nabije Oosten, Leiden.

Uranić, I. (2007). *Aegyptiaca Zagrabiensia: egipatska zbirka Arheološkog muzeja u Zagrebu.* Arheološki muzej u Zagrebu, Zagreb.

de Vartavan, C. (1996). Flore de l'*Égypte* prédynastique (20.000–5.000 BP) : étude préliminaire. *Archéo-Nil*, 6, 9–16.

Vycichl, W. (1972). Sur les noms des parties du corps en égyptiens. *Chronique d'Égypte*, 47(93–94), 173–182.

Walker, J.H. (1996). *Studies in Ancient Egyptian Anatomical Terminology*. Aris & Phillips, Warminster.

Westendorf, W. (1974). Papyrus Berlin 10456: ein Fragment des wiederentdeckten medizinischen Papyrus Rubensohn. In *Festschrift zum 150jährigen Bestehen des Berliner Ägyptischen Museums*, Akademie-Verlag, Berlin, 247–254.

Westendorf, W. (1999). *Handbuch der altägyptischen Medizin*. Brill, Leiden.

WHO (2002). *Preservation and Control of Schistosomiasis and Soil-transmitted Helminthiasis*. World Health Organization, Geneva.

Wilson, J. (1964). Ancient Egyptian medicine. *Journal of the International College of Surgeons*, 41(6), 665–673.

Wreszinski, W. (1912). *Der Londoner medizinische Papyrus (Brit. Museum Nr. 10059) und der Papyrus Hearst: In Transkription, Übersetzung und Kommentar*. J.C. Hinrichs, Leipzig.

Wugh, R.L. (1995). *The Eye and Man in Ancient Egypt*. Wayenborgh, Oostende.

Yamazaki, N. (2003). *Zaubersprüche für Mutter und Kind: Papyrus Berlin 3027*. Achet Verlag, Berlin.

Zaki, M.E. (2013). Success of long bone fracture healing in ancient Egypt: A paleoepidemiological study of the Giza Necropolis skeletons. *Acta Medico-Historica Adriatica*, 11(2), 275–284.

Zink, A.R., Haas, C.J., Reischl, U., Szeimies, U., Nerlich, A.G. (2001). Molecular analysis of skeletal tuberculosis in an ancient Egyptian population. *Journal of Medical Microbiology*, 50(4), 355–366.

Zink, A.R., Grabner, W., Reischl, U., Wolf, H., Nerlich, A.G. (2003). Molecular study on human tuberculosis in three geographically distinct and time delineated populations from ancient Egypt. *Epidemiology & Infection*, 130(2), 239–249.

2

A Retrospective History of Babylonian Medicine

Markham J. GELLER
University College London, UK

The usual surveys of Babylonian medicine follow standard historical chronologies of Mesopotamia, beginning in the early 2nd millennium BCE (the Old Babylonian period) and ending at some point in the Seleucid or Hellenistic centuries of the late 1st millennium BCE, assuming a rather abrupt end to Akkadian medical writings. All of this has been successfully documented[1], although many gaps in our knowledge remain, since medical texts are often fragmentary and not consistently attested in all periods. Nevertheless, the problem with this approach is that it fails to grasp any possible more far-reaching influences of Akkadian medicine on Syriac and Aramaic (and possibly even Arabic) medical knowledge. For this reason, the present chapter argues that the substantial weight of traditional Akkadian medical theory and practice impacted medicine of Late Antiquity.

2.1. Syriac Book of Medicine (3rd century CE)

The very first entry in the third section of the Syriac Book of Medicine is sufficiently diagnostic to indicate the medical background of the recipe. The prescription against migraine consists of a series of separate *simplicia* (a single

1 See, for example, Robson (2008, pp. 455–483).

drug against a single ailment), within the typical Babylonian format of *Listenwissenschaften*[2]:

lnqwš' drš' wplgwt nqwš'

For the throbbing of the head and for the half (of the head), the following steps are used:

dwq ḥdw' wsym lrš'

Pound endive and apply it to the head.

'w ᶜṣr' dqᶜwr' swk bnyḥrwhy mn gb' dnqš

Or: crush extract of date-palm heart in the nostril (Akk. *nahīru*) on the side that throbs.

'w ḥl' wdbš' ḥlwṭ bḥdd' wṭwš ᶜl rš' wqw' qlyl w'šyg bmy'

Or: mix vinegar and honey together and rub on the head and wait (Akk. *waqû*) a little and wash with water.

'w qṭm' wḥl' gbwl bḥdd' wṭwš ᶜl rš'

Or: knead ash (Akk. *qitmu*) and vinegar together and rub on the head.

'w 'wqd gwz' wbṭm' wrws ᶜlyhwn ḥl' wḥlṭ bḥdd' wṭwš ᶜl rš'

Or: burn nut and terebinth (Akk. *buṭumtu*) and sprinkle vinegar over them and mix together and rub on the head.

'w 'pṣ' ḥd' mnt' dwq wgbwl bᶜṭrn' wṭwš

Or: pound one mina of gall-nut and knead in potash (= Akk. *idrānu*) and rub on.

'w šlwq sᶜr' w'rm' bsdwn' wkrwk ᶜl rš' kd šḥynyn wnwḥ

Or: decoct barley and put it on a cloth and you wrap on the head while hot (Akk. *bahrussu taṣammid*)[3], and it will provide relief.

2 Budge (1913, vol. I 553, ll. 5–17), Geller (2024, pp. 11–12).

'w 'wqd kby' dḥmr' wgrm' dsrṭn' wqlyl ḥn' gbwl bmšḥ' wṭwš

Or: burn the dung of an ass (= Akk. *kabūt immeri*) and bones of a crab and a small amount of henna, knead in oil and rub on.

'w mrrt' dqqwn' ᶜm mšḥ' dššm' wṭwš ᶜl rš'

Or: (take) a gall of a *qqwn'*-bird (Akk. *qaqānu*) in oil of sesame (= Akk. *šamaššamū*) and rub on the head.

'w mwḥ' dšq' dymyn' dᶜrb' dl' mlyḥ šḥn qlyl ᶜl nwr' wswk bnḥyrwhy

Or: Lightly heat up over a fire the marrow of the right leg of a ram (Akk. *kurīti immeri*)[4], which is not salted, and smear in his nostrils.

Although these examples have been chosen at random, every recipe has some connection with Akkadian medical ingredients or instructions to an extent that is unlikely to be coincidental. Furthermore, no Greek or Persian loanwords can be detected among these examples, further reinforcing the idea that these Syriac prescriptions were based upon earlier Akkadian models, even if they are no longer available to whoever composed the Syriac. These samples were unlikely to have been directly translated or copied from Akkadian but nevertheless show Akkadian influence and could have been remembered and communicated orally within certain learned circles.

A compound recipe from the same context of the Syriac Book of Medicine shows similar characteristics:[5]

lrš' kgl

For a bald head:

'wqd grm' dmyt' wrwᶜ 'yk qṭm' w'rm' bmšḥ' dzyt' w'rtḥ wṭwš ᶜl rš' wrws ᶜl rš' ḥl' 'w 'yt' ṣypwn' rqy wšmr' wdwq wḥlṭ ᶜm mšḥ' ᶜtyq' dᶜn' wsym tḥwt kwkb' tlt' lylwt' d'rbᶜ bšb' wbtrkn šwp brš'

Burn the bones of the dead and crush (it) like ash (Akk. *qitmu*) and put (it) in olive oil and boil and rub (it) on the head. Sprinkle vinegar on the head.

3 This phrase occurs in an Akk. recipe for migraine; see Scurlock (2014, p. 557, 559).

4 This is a calque on Akk. *eṣemti arki and kurīti immeri*, referring to the leg bone of a sheep.

5 Budge (1913, vol. i 554, ll. 15–19) Geller (2024, pp. 15–16).

> Bring *compressed* soap and fennel (Akk. *šimru*), and pound and mix with old sheep fat (Akk. *šaman immeri labīru*) under the stars for three nights, on the fourth (day) of the week (lit. seven), and afterwards smear (it) on the head.

The recipe calls for the use of human bone as a medical ingredient[6], which is typical of Akkadian *Dreckapotheke,* which often represents secret names (*Decknamen*) for ordinary plants. Additionally, leaving *materia medica* out overnight under the stars or under a specific star was a standard procedure typical of Akkadian prescriptions, as noted by Erica Reiner[7].

2.2. The Babylonian Talmud

In any case, the residual influences of Akkadian on Syriac can be detected only within the applied medicine recipes, with little or no information regarding any theoretical foundations of earlier Babylonian medicine. For instance, the Syriac documentation reveals little about diagnosis or pharmacology, for which we have extensive Akkadian corpora still known from Seleucid Babylonia and earlier, although we cannot know for how long these texts were still being studied. In more specific terms, there is no literature in Aramaic or Syriac devoted to explanatory lists of plants and drugs or descriptions of symptoms, corresponding to the rich Akkadian documentation from earlier. Nevertheless, for some indications of the survival of an Akkadian medical episteme, we can turn to the Babylonian Talmud, which occasionally incorporated useful anecdotes and observations reflecting drug theory and diagnostics:

Shab 109b:

> But one may eat *ywᶜzr*. What is *ywᶜzr*? *pwtnq*. For what (ailment) is it consumed? against *worms* (Akk. *urbatu*). With what (in a recipe) is it consumed? with seven white dates. From what is it (i.e. does it originate)? (From consuming) fatty tissue (Akk. *himṣu*) and water on an empty stomach, or from fatty meat on an empty stomach, or from ox meat on an empty

6 See Scurlock (2014, p. 622, 624), a recipe prescribing rubbing the patient with "human bone" (*eṣemti amēlūti*) in oil, although Scurlock points to this ingredient as a *Deckname* for the plant *šammu ēdu* ("single plant"), for which, see *Chicago Assyrian Dictionary* E, p. 38.

7 See Reiner (1995, pp. 48–50), also drawing attention to Greek *astronomein,* "to expose to stellar radiation". The original Akk. recipe probably specified that the drugs were to be left out under the stars (see Akk. *ina kakkabī tušbat*, "you leave (the drugs) out under the stars") for three nights, and then from days four to seven the head was to be rubbed with the *materia medica*.

stomach; from (eating) nuts on an empty stomach; from fibres of a *thorny*-plant on an empty stomach, and from drinking water afterwards.

'bl 'wkl hw' yw^{c}zr m'y yw^{c}zr pwtnq lm'y 'kly lh l'rbt' bm'y 'kly lh bšb^{c} tmry ḥywwrt' mm'y hwy' m'wmṣ' wmy' 'lyb' ryqn' wmbyśr' šmyn' 'lyb' ryqn' wmbyśr' dtwr' 'lyb' ryqn' m'mgwz' 'lyb' ryqn' wmgydy dkwby' 'lyb' ryqn' wmšty my' 'btryh

There are several noteworthy aspects of this mostly Aramaic passage, which begins with a comment in Hebrew about the plant name *yw^{c}zr,* which may not have been familiar in Babylonia, since the Talmud had to supply an Aramaic gloss (*pwtnq*). This type of glossing of plant names was commonplace within Akkadian schools, which had to cope with plants known in the medical literature on the basis of their Sumerian labels. In this particular case, the plant is thought to be effective against "worms", an Akkadian loanword (*urbatu*). The putative theory in this passage for the presumed cause of the condition – that is, from consuming various foods on an empty stomach – probably represents a misunderstanding of earlier Akkadian medical prescriptions, which regularly recommend that drugs be administered on an empty stomach[8]. In effect, the Babylonian Talmud inherited earlier medical data, which they tried to apply, even though much of the curriculum-based technical knowledge was no longer available.

2.3. Theory in Greek medicine

The interesting question is whether we might be able to map out some of the changes that took place between earlier and later periods in terms of how medical information and practices evolved. For a diachronic view, it is helpful to examine somewhat earlier Greek medical sources from the Hellenistic era for insights into popular medicine within the Mediterranean region. As is well known, after Hippocratic writings became established as a standard text, various new medical philosophies developed within the Greek-speaking world, some aspects of which resembled Akkadian medicine. According to a recent study by Michael Chase, the three pillars of Empiricist medicine were (1) first-person observation of symptoms (*autopsia*), (2) the *historia* of first-person observations of previous doctors and (3) "transition from the similar" (i.e. analogy), which allows the physician "to extrapolate from what he experiences and apply it to previously unfamiliar cases"[9].

8 Such misunderstandings in the Talmud occur regularly since the final redaction of the Talmud took place in the 6th or 7th century CE under obscure circumstances and long after Akkadian was no longer legible.

9 See Chase (2022, pp. 23–48, especially 31–32).

Chase further points out that Empiricists did not follow Aristotelian and other models requiring "a separate faculty of reason" but that Empiricists "rehabilitated the importance of artisanal-technical knowledge" in their approach to medicine. In effect, Empiricists "believed that observation and the faculty of memory, with its tendency to associate phenomena, were all that was needed"[10]. What Michael Chase fails to note are significant parallels with other systems of non-Greek medicine from antiquity, such as the general importance of observation and memory in both older rival systems of Babylonian and Egyptian medicine. Additionally, the so-called Empiricist third pillar of "transition to the similar" is relatively universal in ancient medicine, and treatments for familiar pathologies are also applied to unfamiliar varieties. Even the attribution of medicine to "artisanal-technical knowledge" is typically Babylonian, since the *ummânu* or "expert", the highest rank of scribal school authority, is essentially a craftsman par excellence, according to the basic meaning of this term. What is crucial, however, about Empiricists is their rejection of the pursuit of "hidden causes" for disease, with emphasis instead on pharmaceutical treatments on the basis of observations, which is also typically Babylonian. What appears to have emerged in the Hellenistic period (among both Empiricist and Methodist schools)[11] was a practical approach to medicine that rejected humoral theory and other approaches to medicine on the basis of philosophy rather than empirical data. The influence of these approaches may have been far-reaching, beyond the borders of the Greek-speaking world. A popular approach to ancient medicine that favored practical treatment over theory may have paved the way for many medical systems of Late Antiquity outside the elite confines of Rome and its circle of philosopher-physicians, the most famous of whom being Galen. The movement toward the practical use of recipes for observable symptoms and for the rejection of "hidden causes" may have ultimately resulted in the type of medical systems known in Aramaic and Syriac, in which recipes mainly survive in literary collections of medicine, while more theoretical discussions of medicine became neglected and bereft of any literary heritage.

2.4. Theory in Akkadian medicine

Could this situation have applied to Akkadian medicine as well? In other words, was there a theoretical component to Babylonian medicine that was eventually forgotten, even while technical Akkadian names for drugs and symptoms and the basic formats of Akkadian prescriptions influenced later Aramaic medicine?

10 Ibid., p. 33.

11 See Tecusan (2004, p. 706).

To date, no one has attempted to define in detail any forms of Akkadian medical theory apart from the influences of religion, and one typical approach is to assume that no such theory ever existed in Akkadian; a common assumption is that Greeks indulged in theory, whereas Babylonian scholarship contented itself with empirical lists of data in the form of *Listenwissenschaften*. On the other hand, although "philosophy" as a literary genre never existed in Akkadian intellectual circles, the widespread use of *Listenwissenschaften* provided the means to collect and systematize data in ways that could underpin unstated theoretical propositions. In other words, any theory of medicine would have been implicitly rather than explicitly expressed.

One useful starting point in the search for theory in Babylonian medicine can be found in the area of pharmacology, which required considerable expertise to be able to identify the effects of a drug in relation to any specific symptom. For this reason, the emphasis on *simplicia* in pharmacology is significant in the use of single drugs to counteract a single symptom or pathology. While Akkadian therapeutic texts employed a bewildering pattern of both simple and compound recipes, one text in particular attracts attention as a form of *Listenwissenschaften* devoted to individual *simplicia* in relation to diseases or symptoms. The six-column Vademecum, known as BAM 1[12], provides a list of drugs in a seemingly random order (from our modern perspective) in the first column, associated with diseases or symptoms in the second column, combined with advice for the administration of the drugs in the third column. In effect, BAM 1 provides a minimal record of basic information for treatments of a limited (not exhaustive) list of pathologies. The final column of this unique tablet cites a passage from terrestrial Šumma ālu omens, without explanation, with the only obvious connection being that the omens coincidentally cited many of the same drugs as in the Vademecum. The sample passage from BAM 1 is as shown in Table 2.1.

col. i	col. ii	col. iii
drug *purupuhu*	drug for the sick intestines	to pound, to drink in premium beer
drug root of *kurkanû*	drug to remove *maškadu*-disease	to wrap (in) goat wool, to place on a man's neck
drug shepherd's crook(-plant)	drug ditto	to wrap wool [on a man's neck], to rub on in oil
drug root of thistle	root of [... drug] ditto	to keep rubbing on in oil
drug resin of tamarisk	drug for paralysis	to wrap goat wool to place on a man's neck

Table 2.1. *An extract from BAM 1*

12 This was the very first tablet copied by Franz Köcher in his monumental 6-volume work, *Die babylonisch-assyrische Medizin in Texten und Untersuchungen*. The tablet has been edited by Attia and Buisson (2012) and translated and discussed by Geller (2020).

The point about such *simplicia* is that they may indicate a specific active ingredient, in contrast to ingredients in a compound recipe. What is not possible, however, is to define the perceived nature of these drugs in terms of their usual ancient taxonomies (even in Galen), of "hot" drugs used against "cold" symptoms, or "dry" against "moist", etc., which was a standard ancient aspect of drug theory. Nevertheless, interest in the properties of medicinal plants and minerals can be found in two texts known as Šammu šikinšu (lit. "a plant – its character") and Abnu šikinšu ("a stone – its character")[13].

A sample from the former text indicates how two plants are compared with each other (an example of the "transition to the similar", discussed above in relation to Greek medicine), as well as identifying its medical applications and how it is to be administered. A sample passage reads[14]:

> The plant whose appearance is like (that of) poplar, whose leaves are shiny, whose seed is brownish like [tal]low – [that plant] (2) is called [field]-clod-plant; it is good for stopping *šīqu*-cough. You dry it, (then) he shall drink it at regular intervals in either [wine o]r premium beer [and] he will be cured (transl. Stadhouders 2012).

It is hardly possible from any single entry to establish whether a "theory" of plant lore is involved, especially in the absence of any explanatory data. Nevertheless, this shows that Babylonian scholars or physicians were asking the right sort of questions, perhaps in terms of a dry substance ("field clod") being used against a moist pathology ("cough").

The comparable Abnu šikinšu text follows similar patterns, although any stone that is named is given more general descriptions, for example, compared with a lamb's cranium, a slice of pomegranate, unfermented wine, canal waters or even sunrise[15]. The stone descriptions are less informative than those for plants, since no mention is made of the use of stones against specific medical symptoms, but this

13 The explanatory plant lists known as Šammu šikinšu equate one plant with another plant and then identify for which ailment the plant was useful as a drug (see Stadhouders 2011, 2012). For the corresponding text dealing with stones (Abnu šikinsu), see Schuster-Brandis (2008, pp. 24–47).

14 Stadhouders (2012, p. 2).

15 Schuster-Brandis (2008, p. 37).

may be because semi-precious stones were commonly used as amulets, even in cases of medical therapies[16].

Apart from such explanatory lists, other extensive catalogs of medicinal plants and minerals were collected into various lists, best known under the heading of Uruanna, which remains to be edited and fully contextualized[17]. One unusual feature of this catalog is a section that identifies substances appearing to be *Deckname* or secret names of ordinary plants corresponding to *Dreckapotheke* in therapeutic recipes[18]. Although the topic of *Dreckapotheke* is controversial in terms of how literally to interpret such ingredients within recipes, nevertheless, the listing of such items within Uruanna indicates a level of speculation consistent with the curriculum rather than praxis. Furthermore, many of the drugs listed in Uruanna are either unattested or rarely employed in ordinary prescriptions, which is also indicative of a theoretical or academic approach to pharmacology rather than simple inventories of drugs (which also existed).

2.5. Anatomy

These explanatory lists of plants and minerals are not the only witnesses to forms of academic or theoretical medical literature emanating from the Mesopotamian curriculum. The topic of human anatomy was also of intellectual interest to scholarship, although not necessarily for practical medical applications, as one might expect. One text, known as UGU.MU (literally, "my cranium"), catalogs parts of the human body but not always in the usual *de capite ad calcem* order[19]. The lists appear in Sumerian unilingual and Sumero-Akkadian bilingual formats, identifying parts of the human body never mentioned in medical prescriptions, such as Adam's apple or

---

16 Chains of specific stones were to be worn around the neck as a standard feature of healing practices, perhaps for psychological reasons, and descriptions of these chains feature in elaborate lists of amuletic stones for therapeutic use. These same stones or minerals in powdered form could also appear in salves and potions, to be employed against symptoms thought to be generated by various forms of witchcraft or divine anger, but in any case resulting in illness. Although from a modern perspective such lists would be classified as "magic", it seems clear that the use of amuletic stones belonged as much to the realm of therapeutic medicine as well as to incantations and rituals.

17 A new edition from JoAnn Scurlock is expected to appear soon, while copies of the cuneiform tablets have recently been published by Fincke (2021).

18 See the discussion of Rumor (2017, pp. 1–34).

19 See Landsberger (1967, pp. 51–73), with a general study of the text by Couto (2009). The latest version of UGU.MU with an English translation appears in Civil (2012, pp. 148–162).

nape of the neck[20]. Although it is possible to consider this text to be a theoretical exercise rather than a record of realia, nevertheless, the topic is relevant to medicine, since the final entries in the list refer to physical ailments (boils, inflammation, lesions) as part of the description of the human body[21]. The point is that the topic of human anatomy featured in the curriculum is independent from medical writings, although in neither case reflecting knowledge gained from dissecting human bodies. A similar genre of curriculum text was a list of diseases in both unilingual (Sumerian) and bilingual (Sumerian–Akkadian) formats[22], many of which are not referenced in medical prescriptions. Texts such as these may have provided the basis for academic discourse and oral commentaries, although one cannot determine to what extent they survived into the medical writings of Late Antiquity, or to what extent the academic interests in anatomy and disease may have influenced the practice of medicine.

2.6. Diagnostic

Apart from medical prescriptions, one of the main sources of Akkadian medicine was a 40-tablet composition of diagnostic symptoms, conveniently labeled as the Diagnostic Handbook. This collection of several thousand symptoms, expressed in the traditional form of omens (if P then Q), was intended to provide prognoses for many different diseases to document whether the patient was likely to survive or succumb to his symptoms, or live in chronic poor health. The entries also identified diseases associated with symptoms from various parts of human anatomy, listed in head-to-foot format, providing a wealth of detail unprecedented in any other system of ancient medicine[23]. However, neither the organization of this material nor its omen-format was intended for practical application at the bedside of a patient, since the enumerated symptoms are not based upon individual cases or incidences of disease but as a reference work associating symptoms with possible outcomes. As such, the Diagnostic Handbook can be viewed as a work of theory, using patterns of symptoms as a means of forecasting, approximating the semiotics employed in many other kinds of omens. However, the usual distinctions between "provoked" and "unprovoked" omens (e.g. liver omens vs. earthquakes) rarely apply to medical symptoms or the Diagnostic Handbook. Nevertheless, the conceptual framework of symptoms as prognostic "signs" may have been carried over into later medical

20 See Civil (2012, p. 156).

21 Ibid., p. 158.

22 Landsberger (1967, pp. 77–102).

23 Parallels with Hippocratic prognosis have been highlighted in Labat (1951, p. xxxvii); see Geller (2000–2001, p. 63). For an opposing view, see Asper (2015, pp. 19–46).

thinking, although there is no evidence for the continued existence of the Diagnostic Handbook in the literature. The chief characteristic of the Diagnostic Handbook is its highly systematic structure, providing alternative choices of symptoms for each individual prognosis, such as in the following extract[24]:

> If (the patient's) face is yellow, (then) the Lamaštu-demon has seized him.
>
> If his face is yellow and his eyes are sunken, he will die.
>
> If his face is yellow, the inner part of his eyes is yellow, and the base of his tongue is black, (it is) *ahhazu*-jaundice.
>
> If his face becomes yellow and becomes black, he will die.
>
> If his face is black, his illness will be prolonged and he will (eventually) die.
>
> If his face is black (and) his tongue is red, his illness will be prolonged and he will (eventually) die.
>
> If his face becomes black and becomes white, he will die.

The primary diagnostic colors – yellow, red, black and white – are the same as those associated with humoral theory in Greek medicine, although without reference to bodily humors or fluids in the Diagnostic Handbook. Nevertheless, the lack of any explicit hypothesis in Akkadian does not exclude the possibility of a theory behind these reported observations and prognoses on the basis of the patterns noted in the observations.

One of the significant features of the Diagnostic Handbook is that its contents are summarized in a concise catalog of the incipits of its 40 tablets, comprising one of only three such medical and magical catalogs thus far known from Mesopotamia. The other two cuneiform manuscripts are from the city of Assur, dating from the 8th century, consisting of a catalog of all known medical works listed by incipits, as well as a separate catalog of magical rituals and incantations. All three of these catalogs, now studied in considerable detail in a recent volume[25], were attributed in antiquity to a famous scholar known as Esagil-kīn-apli, who flourished in the 11th century BCE. Like Hippocrates after him, this sage was unlikely to have edited the large corpus of medical and magical texts associated with his name, but the

24 DH Tablet 9, 9–15, see Schmidtchen (2021, p. 415) and Scurlock (2014, pp. 67, 69–70).
25 Steinert (2018).

attribution was sufficiently prestigious for his successors to establish an identifiable corpus or canon of medical, magical and diagnostic texts, which would later be preserved in more-or-less standard recensions. It may not be a coincidence that two of the catalogs of medical and magical texts are known from the city of Assur, which was a key academic center in its day, the Harvard of ancient Mesopotamian science. The importance of these catalogs should not be underestimated, since it now seems likely that the Assur catalog of medical texts influenced the later 7th century BCE Royal Library of Assurbanipal at Nineveh, which incorporated into its vast collections a medical encyclopedia organized according to the Assur medical model[26]. What is impressive in this literature is the existence of a highly systematic wide-ranging corpus of medical literature some two centuries before Hippocrates, not stemming from the distant Orient but from a neighboring society that had much in the way of interaction with Greece and its culture.

One could carry the story further back into the early second millennium BCE, highlighting texts from the time of Hammurabi of Babylon, but the essential picture would remain the same. The oldest examples of such Akkadian medicine from the so-called Old Babylonian period tend to consist of simple recipes with mostly single drugs, without the elaborate addition of incantations and medical procedures (rituals), which became typical of medicine in later epochs. Nevertheless, there are enough congruities to allow us to compare older and later versions of recipes to be able to conclude that the systematic use of drugs was essentially the same, even if more developed over time. For example, a recipe from the early 2nd millennium excavated at Ur reads[27]:

> If a man's eyes have a membrane (*šišītu*), you procuregall of a doormouse ... He shall mix (it) in a green bowl with pure salt.

A similar recipe from the later Nineveh Royal Library reads[28]:

> If a man's eyes are sick, and a membrane (*šišītu*) covers the surface of the pupil of his eyes, tear(s) flow, (and) his eyes are film-filled: *errû*-colocynth, *sahlânu*-plant, seeds of [dog's] tongue plant [...], white plant, bat guano. Those eleven[sic!] drugs you pound [together], mix [in] cedar [oil], daub his eyes, then he should recover.

26 As discovered by Panayotov (2018, pp. 89–120).

27 Geller-Panayotov (2020, pp. 5–6).

28 Ibid., p. 136.

This retrospective view of Mesopotamian medicine raises many questions that have yet to be properly addressed. The existence of a long enduring and wide-ranging system of medicine from Mesopotamia was not only more consistent and homogeneous than its ancient Egyptian counterpart, but it dealt with many of the same essential challenges – prognosis, diagnosis and non-surgical treatment – facing Hippocratic physicians in their own medical milieu. Not following innovative Greek medical theory in the use of diet and eventually phlebotomy, Akkadian medicine adhered to a relatively conservative system of pharmaceutics inherited from earlier periods, which was transferred in its fundamental formats to later Aramaic and Syriac medicine in Late Antiquity. Following these patterns allows us to follow long-term trends in ancient medicine from the Ancient Near East and Mediterranean, resulting in a different view of *Wissenstransfer* from ancient to medieval science from that currently holding sway.

2.7. References

Asper, M (2015). Medical acculturation? Early Greek texts and the question of Near Eastern influence. In *The Frontiers of Ancient Science*, Holmes, B. and Fischer, K.-D. (eds). De Gruyter, Berlin, 19–46.

Attia, A., Buisson, G. (2012). BAM 1 et consorts en transcription. *Journal des Médecines Cunéiformes*, 19, 22–50.

Budge, E.A.W. (1913). *The Syriac Book of Medicines*, volume I. Oxford University Press, London.

Chase, M. (2022). Experience and knowledge among the Greeks: From the Presocratics to Avicenna. In *Premodern Experience of the Natural World in Translation*, Kraus, K., Auxent, M., Weil, D. (eds). Routledge, New York/London, 23–48.

Civil, M. (2012). *The Lexical Texts in the Schoyen Collection (CUSAS 12)*. CDL Press, Bethesda.

Couto, E. (2009). *Etnoanatomía y partonomía del cuerpo humano en sumerio y acadio: El léxico Ugu-mu*. PhD Thesis, Universitat Pompeu Fabra, Barcelona.

Fincke, J.C. (2021). *An Ancient Mesopotamian Herbal Handbook*. Peters, Leuven.

Geller, M.J. (2000). West meets East, early Greek and Babylonian diagnosis. *Archiv für Orientforschung*, 48(49), 50–75.

Geller, M.J. (2020). An apothecary's handbook. *Journal des Médecines Cunéiformes*, 35, 1–33.

Geller, M.J. (2024). *The Syriac Book of Medicines – Section Three – The Local Recipes*, Preprint No. 517. Max Planck Institute for the History of Science, Berlin.

Geller, M.J. and Panayotov, S.V. (2020). *Mesopotamian Eye Disease Texts.* (Babylonisch-assyrische Medizin, 10). De Gruyter, Berlin.

Köcher, F. (1963-1980). *Die babylonisch-assyrische Medizin in Texten und Untersuchungen*, 6 volumes. De Gruyter, Berlin.

Labat, R. (1951). *Traité akkadien de diagnostics et pronostics médicaux*. Brill, Leiden.

Landsberger, B. (1967). *Materials for the Sumerian Lexicon.* Pontificium Institutum Biblicum, 51–73.

Panayotov, S. V. (2018). Notes on the Assur medical catalogue with comparison to the Nineveh medical encyclopaedia. In *Assyrian and Babylonian Scholarly Text Catalogues: Medicine, Magic, and Divination*, Steinert, U. (ed.). De Gruyter, Berlin, 89–120.

Reiner, E. (1995). *Astral Magic in Babylonia.* American Philosophical Society, Philadelphia.

Robson, E. (2008). Mesopotamian medicine and religion: Current debates, new perspectives. *Religion Compass*, 2(4), 455–483.

Rumor, M. (2017). The "AŠ-section" of Uruanna III in *Partitur*. *Journal des Médecines Cunéiformes*, 29, 1–34.

Schmidtchen, E. (2021). *Mesopotamisches Diagnostik. Untersuchungen zu Rekonstruktion, Terminologie und Systematik des babylonisch-assyrischen Diagnosehandbuches und eine Neubearbeitung der Tafeln 3–14.* De Gruyter, Berlin.

Schuster-Brandis, A. (2008). *Steine als Schutz- und Heilmittel, Untersuchung zu ihrer Verwendung in der Beschwörungskunst Mesopotamiens im 1. Jt. v. Chr.* Ugarit-Verlag, Münster.

Scurlock, J. (2014). *Sourcebook for Ancient Mesopotamian Medicine*. SBL, Atlanta.

Stadhouders, H. (2011). The pharmacopoeial handbook Šammu šikinšu – An edition. *Journal des Médecines Cunéiformes*, 18, 3–51.

Stadhouders, H. (2012). The pharmacopoeial handbook Šammu šikinšu – A translation. *Journal des Médecines Cunéiformes*, 19, 1–21.

Steinert, U. (2018). *Assyrian and Babylonian Scholarly Text Catalogues: Medicine, Magic, and Divination.* (Babylonisch-assyrische Medizin, 9). De Gruyter, Berlin.

Tecusan M. (2004). *The Fragments of the Methodists*. Brill, Leiden.

3

Medicine and Healing in Early China

Dolly YANG
Independent Researcher, London, UK

3.1. Introduction

In China, between the late Warring States period (475–221 BCE) and the Han dynasty (202 BCE–220 CE), physicians and scholar-officials adopted and adapted certain philosophical theories, such as *qi*, *yin–yang* and the five agents. This new "medical discourse" enabled them, in an unprecedented way, to map out the human body, understand the causes of disease, diagnose pulses and develop medical treatments such as acupuncture, moxibustion and drug therapy. These medical ideas and practices, naturalistic and holistic in character, initiated a "paradigm shift" that reshaped the entire spectrum of health and healing, laying the foundation for the Chinese medicine we know today. This chapter focuses on the extraordinary development and transformation of medical theory and practice that took place between the 3rd century BCE and the 3rd century CE, namely, from the end of the Warring States period to the end of the Han dynasty. It begins with a brief outline of the social and political backdrop from which the new medical paradigm emerged. The core of the chapter is divided into the following six themes:

1) Notations of microcosm-macrocosm.

2) Formation of canonical texts.

3) Classical theory.

4) Diagnosis.

History of Life Sciences and Medicine 1,
coordinated by Lorenzo PERILLI.

5) Therapy.

6) Pharmacopeia.

Several groundbreaking works by medical historians, such as Donald Harper, Li Jianmin, Nathan Sivin, Paul Unschuld, Vivienne Lo and Yamada Keiji, have illuminated our understanding of this foundational period of Chinese medicine. Drawing on current historical scholarship, this chapter highlights the pluralistic nature of medical theory and practice, which became increasingly systematized and standardized toward the end of the Han dynasty. It also recognizes that the concept of "medicine" in this period, relating as it does to a whole gamut of self-cultivation practices later known as *yangsheng*, is much broader than "what physicians did".

3.2. Medicine during the late Warring States, Qin and Han dynasties

The Warring States period (475–221 BCE), the last two centuries of the declining Zhou dynasty, was a turbulent time, with constant warfare between seven great states, each vying with the others for overall hegemony. It was also a time of great cultural and intellectual expansion. Many prominent thinkers emerged during this period of social and political turmoil, who were very much concerned with the art of rulership and statecraft. An increasingly wide array of specialists, described by later Han writers as *fangshi* (masters of formulas), also began to receive attention and patronage for their expertise in "natural philosophy and occult knowledge", such as calendrics, astrology, divination and medicine (Harper 1998, 1999). These specialists expounded the theories of *qi*, *yin* and *yang* and the five agents (*wuxing*), which, by the end of the 3rd century BCE, coalesced into a system of "correlative thinking" with elaborate correspondences between the cosmos, the state and the body (Graham 1989, pp. 313–370; Sivin 1995).

While *fangshi* is an eclectic term referring to a range of specialists in the technical arts who shared a common interest in understanding natural phenomena, *yi* is a more specific title for physicians during the late Warring States, Qin and Han periods. These physicians developed new medical theories using a set of technical languages, such as *qi*, *yin–yang* and the five agents, and treated their patients according to the new conception of the body as a microcosm of the cosmos. In doing so, they distinguished themselves from other healers, such as the *wu*-spirit medium, who still operated within a belief system that considered evil spirits and ancestors to be the sole agents of disease. The Chinese character for "healer" (*yi*) began to change, replacing the exorcistical *wu*-spirit medium in *yi* 毉 with medicinal wine in *yi* 醫. The same character is still used today to signify medicine or physician. This indicates that the practice of medicine was no longer confined to the incantations of the *wu* but was taken over by *fangshi* who treated their patients with wine and other methods. This new medical paradigm, described as the "medicine of systematic correspondence", can be seen as reflecting ideas and sociopolitical structures that emerged from efforts to overcome the chaos of the Warring States and from the subsequent conditions accompanying the first unification of China (Unschuld 1985, p. 51).

In 221 BCE, the ruler of Qin state in the northwest, posthumously known as Qinshi Huangdi (260–210 BCE), finally defeated all the other states and became the first emperor of China. To integrate the formally separate political entities into a unified and centralized political state, a completely new state structure was created, with different parts of the country being linked by a massive system of roads and waterways and by a whole range of reforms, such as standardized currency, weights, measures and writing. The Qin dynasty (221–206 BCE) lasted only 15 years before being taken over by Liu Bang (256–195 BCE), who led a peasant revolt against the Qin emperor and later established the Han dynasty in 202 BCE. The Han dynasty (202 BCE–220 CE) is divided into the Western or Former Han (202 BCE–8 CE) and the Eastern or Later Han (25–220 CE). It was briefly interrupted by Wang Mang (45 BCE–23 CE), who was believed by many to be the usurper who established the Xin dynasty (9–25 CE). Thus, apart from the short interval between 9 and 25 CE, there was a long period of social stability and economic prosperity during this time, which proved to be a crucial watershed for the formation of canonical works in medicine.

During the Qin and Han periods, diverse medical ideas circulated among interconnected cultural centers, such as Qi (in present-day Shangdong), Chu (in present-day Hunan) and Bashu (in present-day Sichuan). As China became unified and centralized, there was a strong impetus to move from numerous parallel medical theories toward a unified and systematic approach, supported by a political bureaucracy striving to conform to a particular standard. Several important medical texts were compiled during the Han period. *Huangdi neijing* (The Inner Canon of the Yellow Emperor ca. 1st century CE, hereafter *Neijing*), the oldest received classic of medical theory, epitomizes the collected knowledge of Han physicians writing within the new medical paradigm.

The relatively homogenized form of the medical system reflected in the received canons of this period underwent a "reality check" in the 1970s and 1980s with the discovery of medical manuscripts and artifacts in various Han tombs, such as Mawangdui (closed 168 BCE), Zhangjiashan (closed ca. 186 BCE), Mianyang (before 118 BCE), and Wuwei (ca. 1st century CE). The most recently excavated medical texts were from the Laoguanshan tombs near Chengdu in Sichuan Province from 2012–13. The theoretical understanding and therapeutic techniques described in these bamboo and silk manuscripts are noticeably heterogeneous and less sophisticated (Keegan 1988; Harper 1998). There are writings on magical therapies, drug-based therapies, breathing exercises, therapeutic exercises, known as *daoyin*, and sexual techniques. Spirit healing and belief in the wrath of ancestors and the malevolence of demons were recorded, forming the pluralistic medical landscape of the time. As many as 56 recipes from various Mawangdui medical manuscripts, for example, contain "magical techniques", including incantations, ritual acts and charms (Harper 1998, pp. 148–183). They also revealed that ancestral and demonological concepts of illness had begun to yield their dominant position to the new medical paradigm whereby the human body becomes a microcosm of the cosmos.

3.3. Notions of microcosm-macrocosm

During the 3rd century BCE, as the political order changed fundamentally, representations of the body, the cosmos and the state reflected new ideas of a correspondence between microcosm (the body) and macrocosm (the cosmos) (Sivin 1995). Zou Yan (ca. 305–240 BCE) was reputed to have developed a theory of cosmogony, using *yin–yang* and the five agents to explain cyclical changes in nature and to predict the political fortunes of rulers. His theory was employed as a theoretical framework in many areas, including ritual, astrology, divination and medicine.

Lüshi chunqiu (The Springs and Autumns of Master Lü), compiled under the patronage of the Qin Chancellor Lü Buwei (291?–235 BCE), who was critical of the rise of the Qinshi Huangdi, expounds on the notion of microcosm and macrocosm as being a single manifold through the concept of *qi*, linking it to *yin–yang* and the five agents (Lloyd and Sivin 2002, p. 264). It sets out a philosophy of statecraft that exalts the ruler, who holds the pivotal position between heaven and earth, as the exemplar of humanity. It is through the ruler's self-cultivation that the harmony of the state can be established. The text transforms the soon-to-be emperor from a conqueror to a life-giver (Sivin 1995).

However, the legalist ideology of law and punishment was the guiding principle of the Qin government, Confucianism, a school of thought based on the works of Confucius (551–479 BCE), which became the orthodox political doctrine during the Han. A harmonious society, according to Confucians, can be achieved by binding individuals to specific social roles and regulating the relationships between these roles through a hierarchical nexus of mutual obligations bounded by customs and rites. In Confucianism, an individual's behavior is guided by five fundamental relationships: between the ruler and subject, father and son, husband and wife, elder brother and younger brother, and friends. During the Han, when "a hundred schools were abolished, honoring only the Confucians", the sociopolitical order of Confucian doctrines was incorporated with the ideas and healing expertise of the *fangshi* into a theoretical system that established the fundamental nature and characteristics of Chinese medicine (Liao 2005). The notion of *tianren heyi* (harmonious unity between the cosmos and humanity) expounded by Confucians is expressed closely as *tianren xiangying* (mutual responsiveness between the cosmos and humanity) encapsulated in *Neijing*, in which the cosmos and people are two complementary aspects of the same phenomenon. *Lingshu* (spiritual pivot), which forms part of the *Neijing* corpus, famously records the following:

> Heaven is round, Earth is square; people's heads are round, their feet are square and thereby correspond to them. Heaven has the sun and moon, people have two eyes; Earth has nine regions, people have nine orifices. Heaven has wind and rain; people have joy and anger. Heaven has thunder and lightning; people have tones and sounds. Heaven has four seasons; people have four limbs. Heaven has five tones; people have five *zang*-organs. Heaven has six pitches, people have six *fu*-organs (*Lingshu* 71).

This passage lays out a series of correspondences between the body and cosmos. While various medical writings during the Han period were based on the same philosophical and political ideology of microcosm-macrocosm relations, their authors often employed the literary device of using an authoritative figure, such as Huangdi (Yellow Emperor), mytho-historical sovereign and cultural hero, Shen Nong (Divine Farmer), another mythological ruler and Bian Que, legendary physician of the Warring States period, as their mouthpiece to record their own medical expositions. Some of these writings became part of the canonical works of Chinese medicine.

3.4. Formation of canonical texts

In Chinese medicine, "canonical" refers to those works known as *jing* (canon), such as the *Huangdi neijing* (Inner Canon of the Yellow Emperor), *Nanjing* (Classic of Difficulties) and *Shennong bencaojing* (Divine Farmer's Classic of *Materia Medica*). There has long been a consensus that these canonical works were produced during the Warring States period, laying the theoretical foundation for physiology, pathology and pharmacology. However, since the discoveries of medical manuscripts and artifacts from various Han tombs excavated in the last 50 years, modern scholars have cast doubts on these dates, arguing that they were produced at a later date, during the Han period, and postdated the material found in the Han tombs (Yamada 1979; Sivin 1993; Liao 2005; Lo 2013).

The first canonical work representing a significantly developed doctrinal framework for medical practice was the *Neijing*. This text sets out the cosmological, metaphysical and medical doctrines, including spiritual and moral principles, which constitute the foundation of Chinese medical practice. It comprised many smaller texts, with different authors, whose material had been shaped and reshaped through a long and active textual tradition. More than half of its material dates from between the 3rd century BCE and 260 CE (Keegan 1988, p. 17). The modern version of the *Neijing*, based on a heavily revised edition from the Song dynasty (960–1279), comprises two independent texts: *Suwen* (Plain Question) and *Lingshu* (Numinous Pivot). The two texts have always been listed separately in the bibliographies of dynastic histories. It was pure speculation on the part of Huangfu Mi (215–282 CE) to assume that these two texts must be the *Neijing*, simply because the two texts combined totaled 18 scrolls, which accords with the "Yellow Emperor's Inner Classic, 18 scrolls" in the bibliographic treatise of the *Hanshu* (Book of Han, around 82 CE) (Liao 1988). Huangfu Mi's work is the earliest surviving compilation of the *Neijing* corpus under the title of *Zhenjiu jiayi jing* (AB Canon of Acupuncture and Moxibustion). This text, which exerted great influence on the clinical practices of acupuncture and moxibustion, became part of the canonical works of Chinese medicine in its own right.

The *Nanjing*, or to give it its full title *Huangdi bashiyi nanjing* (Yellow Emperor's Canon on Eighty-One Difficult Issues), is now generally considered to have been compiled during the 1st or 2nd century CE. The 81 questions asked in *Nanjing* concern pulses, channels, *zang* and *fu* organs, diseases, acupoints, acupuncture and moxibustion. However, *Neijing* discussed *yin–yang* and the five agents separately, *Nanjing* modified the independent theories of *yin–yang* and the five agents of *Neijing* by integrating them into one theory (Unschuld 1986, p. 147). To solve the problem of the discrepancy between the *yin* and *yang* properties of the five *zang* and the six *fu* organs described in *Neijing*, the author of *Nanjing* counted the "kidney" as two organs: "kidney" on the left and "gate of vital force" on the right, thus making six *zang* organs, while claiming that one of the six *fu* organs, *sanjiao* (the triple burner), exists only in name and has no physical form (*Nanjing* 25, 36, 39). Unschuld, who translated *Nanjing* into English, concludes that *Nanjing* was compiled "to overcome the heterogeneity and unsystematic nature of the *Huangdi neijing* anthology of medical schools and concepts" (1986, p. 34). Thus, in *Nanjing*, a great effort has been made to consolidate various aspects of theoretical and practical knowledge into a more coherent and systematic doctrine.

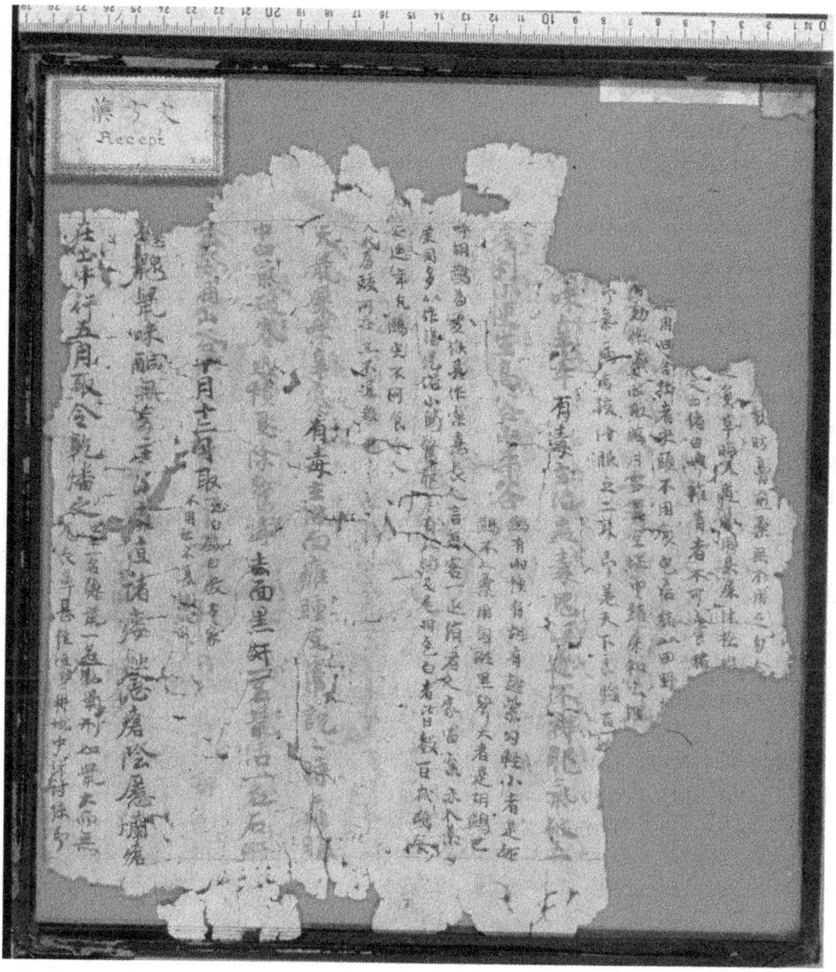

Figure 3.1. *Fragment of Tao Hongjing's Bencao jingji zhu. Courtesy of the Deposit of the Berlin–Brandenburg Academy of Sciences and Humanities, State Library of Berlin, Prussian Cultural Heritage, Oriental Department, Turfan Collection Ch. 1036. For a color version of this figure, see www.iste.co.uk/perilli/history1.zip*

Shennong bencao jing (Divine Farmer's Classic of *Materia Medica*), compiled during the Eastern Han dynasty (25–220 CE), is the first pharmacological treatise in China. This was attributed to Shen Nong, a legendary ruler and ancestral deity who became the patron of the pharmacological arts. Lost not long after its completion, this text survived thanks to 5th-century Daoist master Tao Hongjing (456–536 CE), who incorporated it into his work, *Bencao jingjizhu* (Collected Annotations on the

"Classic of *Materia Medica*"). In *Bencao jingjizhu*, Tao faithfully copied the information from *Shen Nong bencao jing* in large red characters, that from other sources in large black characters, and wrote his own commentaries in small black characters (Figure 3.1). This multilayer format became standard practice for commentary writing in Chinese *materia medica*. In Tao's work, the original *Shennong bencao jing*, comprising three *juan* (chapters), contains descriptions of 365 drugs in three categories: upper-, middle- and lower-class drugs. Each drug's entry lists nomenclature, properties, compatibilities and medical applications. Some include descriptions of production regions and the ecological environment (Unschuld 1986; Wilms 2016).

Although *Shanghan zabing lun* (Treatise on Cold Diseases and Miscellaneous Diseases) lacks *jing* (canon) as part of its title, it also, over time, came to be regarded as part of the canonical works of Chinese medicine. The text, written by Zhang Zhongjing (ca. 150–219 CE) of the Eastern Han dynasty, was lost soon after his death. Fortunately, the section on "cold diseases" was collected by Wang Shuhe (ca. 180–270 CE), an imperial physician of the Western Jin dynasty (265–316 CE), and renamed *Shanghan lun* (Treatise on Cold Disease). In *Shanghan lun*, Zhang Zhongjing discussed the diagnosis and treatment of exogenous disease, that is, cold disease, in terms of the six stages of disease: *tai yang* (greater *yang*), *yang ming* (bright *yang*), *shao yang* (lesser *yang*), *tai yin* (greater *yin*), *shao yin* (lesser *yin*) and *jue yin* (terminal *yin*). This method of demarcating the development of an illness via a threefold *yin–yang* schema, from initial infection to death, has dominated clinical Chinese medicine ever since.

Maijing (Pulse Classic), a treatise on the pulse attributed to Wang Shuhe, became the canonical work of pulse diagnosis. Drawing material on pulses from texts such as *Neijing* and *Nanjing*, *Maijing* contains descriptions of 24 pulse diagnoses that are a "dense tangled mesh of interrelated, interpenetrating sensations", which potentially causes more confusion than clarity (Kuriyama 1999, p. 94). Nevertheless, the distinct method of pulse diagnosis formulated and described by Wang Shuhe in this text exerted a lasting influence on the development of sphygmology.

All these canonical works were based on theories of systematic correspondence, such as *qi*, *yin–yang* and the five agents. These ideas were further developed into other theories specifically for health and medicine, such as the theory of the four seasons, the theory of viscera and their manifestations and the theory of channels and collaterals. Together, they formed a solid conceptual framework for the unique healthcare system that we recognize today as traditional Chinese medicine (TCM).

3.5. Classical theory

Physicians of the late Warring States, Qin and Han periods, who shared the sociopolitical and intellectual milieu of *fangshi*, or technical experts, adopted and adapted various philosophical and political ideas into their practices. They applied ideas of "correlative cosmology", further developing them into the "medicine of systematic correspondence". Their main theories are outlined below.

3.5.1. *Theory of qi*

The term *qi*, depicted in several ancient scripts and carrying a variety of meanings, underwent a process of evolution, with its cosmological dimensions coming into play around 400 BCE (Stanley–Baker 2022). In *Guanzi*, a set of writings on statecraft composed in the late 4th century BCE, *qi* is defined as "what fills the body" (*Guanzi* 37). According to *Zhuangzi*, another Warring States text attributed to the philosopher Zhuangzi (370–387 BCE), "The birth of a human being is an agglomeration of *qi*. While it agglomerates, there is life; when it disperses, there is death" (*Zhuangzi* 22). Thus, by the Warring States period, *qi* had come to be understood as the fundamental substance permeating both the universe and the body. It had become "the backbone of Han cosmology and physical thought" (Sivin 2000, pp. 123–124). In this new concept of the human body, *qi* is the "glue" that unites philosophical, political and medical discourse (Stanley-Baker 2022). The agents of illness, for example, became *xieqi* (evil *qi*) instead of *xiegui* (evil demons). As *qi* was further conceptualized as the interplay between *yin* and *yang*, the occurrence of disorders was attributed to disharmony between the body's *yin qi* and *yang qi* rather than to the wrath of ancestors and demons, as had been assumed in the Shang and Zhou dynasties (ca. 1600–256 BCE).

3.5.2. *Theory of yin–yang*

Yin and *yang* were terms that were originally used to explain the positions of opposing or facing the sun. The place opposed to the sun was *yin* (lit. shadow), while the place facing the sun was *yang* (lit. sun). This basic idea became increasingly elaborate, referring to a wide range of objects and phenomena. The alternation of *yin* and *yang*, through the movement of *qi*, was understood to be a universal law governing the natural world, the state and people.

The use of the *yin–yang* concept as a tool to map the human body and explain the cause of disease can already be found in the excavated Han medical manuscripts. Both the Mawangdui and Zhangjiashan tomb texts give descriptions of 11 *yin* and *yang* channels (Lo 1998, pp. 336–339; Harper 1998, pp. 192–212). Together, they provide the earliest textual evidence of channels, albeit differing from the 12 channels described in *Neijing*.

In *Neijing*, not only were the 12 channels divided into *yin* and *yang*, but all organs and body parts were conceptualized according to the *yin–yang* theory:

> Speaking of the *yin–yang* of a person, then the outside is *yang*, the inside is *yin*.
>
> Speaking of the *yin–yang* of the body, then the back is *yang*, the abdomen is *yin*. Speaking of the *yin–yang* of the *zang*- and *fu*-organs of the body, then the *zang* organs are *yin* and the *fu* organs are *yang* (*Suwen* 4; based on Unschuld and Tessenow's translation, 2011, p. 89).

The theory of *yin–yang* was used extensively to explain the activity of the human body and the cause of diseases, and thus assisted physicians in their diagnosis and treatment strategies. It also underwent twofold, threefold and sixfold differentiations, conceptualized in areas such as the four seasons, Zhang Zhongjing's "six stages of disease" and the 12 channels described in *Neijing*.

3.5.3. *Theory of the four seasons*

At some point, *yin–yang* was further divided into *shaoyang* (mild *yang*), *taiyang* (extreme *yang*), *shaoyin* (mild *yin*) and *taiyin* (extreme *yin*), signifying the properties of the four seasons. The *Neijing* writers emphasize the importance of being in harmony with the four seasons and adopting a daily regimen in accordance with the *qi* of the natural cycles:

> The three months of spring, they denote effusion and spreading. Heaven and earth together generate life; the myriad beings flourish. Go to rest at night and rise early. Move through the courtyard with long strides. Dishevel the hair and relax the physical appearance, thereby causing the mind [to orient itself on] life. Give life and do not kill. Give and do not take. Reward and do not punish. This is in correspondence with the *qi* of spring and it is the Way to nourish life (*Suwen* 2; Unschuld and Tessenow 2011, pp. 45–46).

The emphasis on patterning one's lifestyle to nature's cycle also features in *Yinshu* (The Book of Pulling), a text excavated from the Zhangjiashan Han tomb. It describes the kind of regimen one should follow according to the season and offers a comprehensive step-by-step guide to bodily movement as part of a seasonal health regime (Lo 1998, 2014). A culture of self-cultivation practices, later known as *yangsheng* (Nourishing Life), including breathing exercises, *daoyin* (lit. guiding and pulling), diet, sexual hygiene and other preventative health practices became prevalent among Han nobles (Yang 2018, pp. 39–46). Descriptions of these self-cultivation practices formed an important part of the medical literature of the Han period, as reflected in the *fangji* (recipes and techniques) section of the bibliographic treatise in the *Book of Han* (Harper 1998, p. 6).

The theory of the four seasons was also used to guide diagnosis and explain the etiology of diseases:

> If one examines the [patient] in winter, the [movement in the] vessels on the right must be in the depth and tight under the regular [circumstances]. This is in correspondence with the four seasons. When the [movement in the] vessels on the left is at the surface and retarded, this is contrary to the [course of the] four seasons… As for 'assess', one finds the location of the disease and assesses it on the basis of the four seasons (*Suwen* 46; Unschuld and Tessenow 2011, p. 680).

Although the theory of the four seasons is less well known than those of *yin–yang* and the five agents, it was nevertheless a fundamental building block of the new "medicine of systematic correspondence", stressing the importance of prevention over treatment by sustaining a daily regimen that is in accordance with the rhythm of the seasons.

3.5.4. *Theory of the five agents*

The theory of the five agents seems to have been incorporated into medicine later than those of *yin–yang* and the four seasons. It was used primarily to classify things into five specific categories of wood, fire, earth, metal and water, and to explain their mutual relationships.

Unlike the four elements of the ancient Greeks, the five agents are ever-changing material forces that were constructed into various orders, including a sequence of "mutual conquest" (*xiangke*) and "mutual production" (*xiangsheng*) (Figure 3.2).

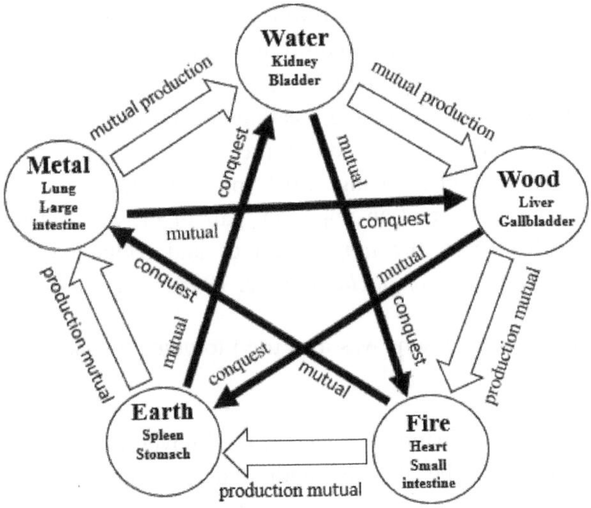

Figure 3.2. *Diagram of the five agents in the sequences of "mutual conquest" and "mutual production", with their corresponding internal organs*

Reference to the five agents is largely absent from the available excavated Han manuscripts, apart from *Taichanshu* (Book of the Generation of the Fetus), a Mawangdui medical manuscript (Harper 1998, p. 79) and the Laoguanshan medical manuscripts and figurine (Lo 2024).

In *Neijing*, a wide range of phenomena, including directions, environmental conditions, colors, flavors, the five *zang* organs and their corresponding body parts (Table 3.1), has been systematically made to correspond to the five agents. These connections play an important role in diagnosis.

Five agents	Direction	Environmental quality	Color	Flavor	*Zang*-organ	Body part
Wood	East	Wind	Green	Sour	Liver	Sinew
Fire	South	Heat	Red	Bitter	Heart	Blood
Earth	Center	Dampness	Yellow	Sweet	Spleen	Flesh
Metal	West	Dryness	White	Pungent	Lung	Skin and hair
Water	North	Cold	Black	Savory	Kidney	Bone marrow

Table 3.1. *Selected correspondence between the five agents and phenomena in Neijing*

The theory of the five agents can be applied in treatments. For example, the 69th issue of *Nanjing* recommends the following:

> In the case of depletions, fill the respective [conduit's] mother. In the case of repletion, drain the respective [conduit's] child (Unschuld 1985, p. 583).

The reasoning here adheres to the "mutual production" sequence. For example, kidney water generates liver wood; hence, the kidney is the mother, and the liver is the child. If there are signs of deficiency in the liver, the recommended treatment is to nourish the kidney water (i.e. the mother) to eliminate the deficiency of the liver. On the other hand, liver wood generates heart fire; hence, the liver is the mother, and the heart is the child. If there are signs of excess in the liver, the recommended treatment is to purge the heart fire (i.e. the child).

An alternative treatment method, which follows the "mutual conquest" sequence, is to prevent further development of a disease, and stop it moving from one organ to the next:

> When one sees an illness in the liver, one should know that the liver will transmit it to the spleen. Hence, one prevents this [transmission] by filling the influences of the spleen, with the effect that it will not accept the evil [influence] from the liver. Hence [the scripture] speaks of 'treating what is not yet ill (*Nanjing* 77; Unschuld 1985, p. 630).

The five-agent theory was adapted into the new medical paradigm of the Han as a classification scheme, a way of explaining the physiology and pathology of the body, as well as facilitating clinical diagnosis and treatment.

3.5.5. *Theory of viscera and their manifestations*

Another important component of the newly emerging "medicine of systematic correspondence" was the theory of viscera and their manifestations, which is concerned with the physiological functions of, and pathological changes in, the internal organs. Individual organs had been known for a long time, but it was not until the 2nd century BCE that they were envisaged as "[growing] together producing an organism in which each organ was dependent on the others and contributed in various ways to supplying the needs of the body as a whole" (Unschuld 1998, p. 20).

In *Neijing*, the internal organs of the body comprise five *zang* (liver, heart, spleen, lung and kidney), six *fu* (gallbladder, stomach, large intestine, small intestine, bladder and triple burner) and another six so-called "extraordinary *fu* organs" (brain, marrow, bones, vessels, gallbladder and uterus). It considers the five *zang* as *yin* organs because they "store essence but do not discharge it"; the six *fu* as *yang* organs because they "transport and transform food without storing it up"; and the extraordinary *fu* organs as *yin* organs because they "are generated by the *qi* of the earth; and their storing is associated with *yin*" (*Suwen* 11; Unschuld and Tessenow 2011, pp. 203–204).

When these organs were incorporated into the classification scheme of the five agents, the five *zang* organs were paired with the first five *fu* organs, leaving out the sixth *fu* organ, that is, the triple burner. The author of *Nanjing*, considering this drawback, invented another *zang* organ, the *mingmen* (gate of vital force), to pair with the triple burner. This sixth pairing created much confusion, not only because *mingmen* refers to the eyes in *Neijing* (*Lingshu* 5.2) but also because, in *Neijing*, the channel of the triple burner had already been paired with the channel of the pericardium (*xinbao*), often referred to as the heart protector, in the formation of the twelve channels. Scholars of later generations tried without success to prove that the pericardium and the "gate of vital force" were identical (Sivin 1987, pp. 126–127).

Elsewhere in *Neijing*, the "twelve viscera" are described as consisting of the five *zang* and six *fu* organs, plus another "organ" called *danzhong*. This region is situated midway between the two nipples and is often interpreted as the pericardium. Using the metaphorical language of a state's bureaucratic functioning, each of the twelve viscera was assigned a variety of tasks, just as officials oversaw administrative duties:

> The heart is the official functioning as ruler. Spirit brilliance originates in it. The lung is the official functioning as chancellor and mentor. Order and moderation originate in it. The liver is the official functioning as general. Planning and deliberation originate in it.… All these twelve officials must not lose [contact with] each other (*Suwen* 8; Unschuld and Tessenow 2011, pp. 155–158).

These internal organs, applicable primarily to the five *zang* and their corresponding *fu* organs, were linked with body parts, orifices, complexion, voice, movements and emotional states, all of which were open to observation. Table 3.2 demonstrates some of the interconnections between the five *zang* organs and their manifestations from *Neijing*.

Zang organ	Its effulgence	Its fullness manifests in	Complexion	Voice	Movement	Orifice	State of mind
Liver	Nails	Sinews	Green	Shouting	Clenching	Eye	Anger
Heart	Face	Blood vessel	Red	Laughing	Belching	Tongue	Overjoy
Spleen	Muscle	Lips	Yellow	Singing	Hiccup	Mouth	Pensiveness
Lung	Body hair	Skin	White	Weeping	Coughing	Nose	Anxiety
Kidney	Hair	Bones	Black	Groaning	Shivering	Ears	Fear

Table 3.2. *The correspondence between the five zang organs and their manifestations in Neijing*

The theory of viscera and their manifestations assists in diagnosis and treatment, with emphasis on the functions of the *zang–fu* organs rather than their anatomical structure. Although indications of anatomical familiarity with the internal organs were already evident in *Neijing*, particularly in *Nanjing*, where detailed measurements of the weight, diameter and length of the liver, heart, spleen, lung, kidneys, gallbladder, stomach, large intestine, small intestines and bladder are recorded, no anatomical approaches seem to have been pursued seriously in China (*Nanjing* 42; Unschuld 1986, pp. 416–417). The understanding of the structure and function of the body was guided by the new structural social and economic environment that accompanied the unification of China.

3.5.6. *Theory of channels and collaterals*

The concept of *jing*, translated here as a channel, follows the metaphor of the natural world's rivers and lakes. However, the word originally referred to the warp threads of a loom. The presence of weaving looms in the Laoguanshan tomb, excavated in 2012–2013, initiated a new understanding of the early emergence of the concept of "channels". They are described in excavated medical manuscripts as warp threads stretched between the top and bottom cross beams of the loom (Lo 2024). These channels were not "linked up", nor did they connect to the internal organs, nor were they described with any points for needling; instead, they ran on the surface of the body and the limbs (Lo 2013, p. 41). For example, the first of the eleven channels described in *Maishu* (The Channel Book) from the Zhangjiashan Han tomb is the "Great *Yang* Channel":

> The Great *Yang* Channel is tied to the heel at the centre of the outer malleolus. It comes out at the centre of the popliteal crease and, rising, it pierces the buttock, coming out on the hip to press on either side of the spine. It comes out at the nape of the neck, rises the corner of the head and descends between the eyebrows and eyes to press on either side of the nasal bone and is tied to the inner edge of the eye (Lo 1998, p. 336).

By the time *Neijing* was compiled (around the 1st century CE), the concept of *jing* had shifted. It was now compared with "stream water", as described in *Lingshu* 71:

> The earth has high mountains; man has shoulders and knees. The earth has deep valleys; man has armpits and the hollows of the knees. The earth has twelve stream waters; man has twelve stream/conduit vessels. The earth has spring vessels; man has guard *qi* (Unschuld 2016, p. 635).

The 12 channels, as the main pathways of the body, are equivalent to the waterways of the unified empire of the Han dynasty. In addition to the 12 channels, several subsidiary channels are found in *Neijing*, including the following:

– The eight extraordinary channels (*qijing bamai*).

– The 12 channels of the tendons (*jingjin*).

– The 12 channels of diversions (*jingbie*).

– The 15 collaterals (*lomai*).

The channels are longitudinal pathways, whereas the collaterals transversely connecting the channels subdivide them. They are passages through which blood and *qi* flow to nourish *yin* and *yang*, moisten sinews and bones, and lubricate joints (*Lingshu* 47).

Two types of *qi* are transported by channels and collaterals: *ying* (lit. camp, also translated as "nutritive" or "supply") *qi*, which brings nutrients to the body and limbs, and *wei* (lit. guard or defense) *qi*, which protects the exterior of the body from pathogenic invasion. The two terms have a strong military flavor, conceptualizing the body as a state protecting and defending its own territory. Pathologically, channels and collaterals are the pathways through which exogenous pathogens travel to different parts of the body. This was later fully theorized in Zhang Zhongjing's "six stages of disease".

The idea of blood and *qi* circulating through the channels emerged from the unique circumstances obtained at the time of the unification of the empire. Only then was it suggested that blood and *qi* move from one channel to another like a circle without beginning and ending. For example, *Lingshu* describes the circular movement of the camp *qi,* which moves from, and back to, the channel of the lung via those of the large intestine, stomach, spleen, heart, small intestine, bladder, kidney, pericardium, triple burner, gallbladder and liver (*Lingshu* 16). The notion of circulating blood and *qi* would have been derived from analogical reasoning rather than anatomical experiments.

All the *zang–fu* organs are linked to different parts of the body and the limbs through channels and collaterals. Thus, a disease of the internal organs can manifest itself on the body's surface, having been transmitted through the channels and collaterals, and can be detected by physicians in their diagnosis.

3.6. Diagnosis

The theory that "what is within must manifest itself without", articulated by Mencius (372–289 BCE, *Mencius*, *Gaozi* 12.6), a Confucian philosopher of the late Warring States period, was elaborated in *Neijing,* as shown in this passage from *Suwen*:

> In the case of heat disease in the liver, the left cheek becomes red first.
>
> In the case of heat disease in the heart, the forehead becomes red first.
>
> In the case of heat disease in the spleen, the nose becomes red first.
>
> In the case of heat disease in the lung, the right cheek becomes red first.
>
> In the case of heat disease in the kidneys, the chin becomes red first.
>
> (*Suwen* 32; Unschuld and Tessenow 2011, p. 506)

The orientation of the five *zang* organs here is based on the theory of the five agents. This was one of many diagnostic techniques described in *Neijing*. The author of *Nanjing* grouped the various diagnostic approaches found in *Neijing* under four diagnostic methods, namely, "looking", "listening", "asking" and "feeling the *mai*" (*Nanjing* 61; Unschuld 1986, p. 539). This became standard practice in Chinese medicine. The word *mai*, often translated as vessels or channels, refers to "pulse" in the context of diagnosis. Thus, "feeling the *mai*" is to examine the movement of the channels in the body that are felt as "pulses".

The earliest extensive account of pulse diagnosis is found in the biography of Chunyu Yi (fl. 176 BCE) in *Shiji* (Records of the Grand Historian, ca. 86 BCE). He was an expert in pulse diagnosis who "received texts on pulse attributed to the Yellow Emperor and Bian Que" and was able to "examine illness by means of the five colors and recognize whether a person would die or live" (*Shiji* 105). The form of pulse diagnosis, in fact prognosis, that Chunyu Yi practiced was to examine the *mai* on the body surface to identify different qualities of *qi* coming from the internal organs (Hsu 2010, p. 4).

Prominent among several pulse diagnostic techniques mentioned in *Neijing* is the "three sections and nine indicators" (*sanbu jiuhou*), with a whole chapter devoted to it (*Suwen* 20). *Sanbu jiuhou* divides the body into three sections (upper, middle and lower). In each section, there are three indicators of locations described as "heaven", "earth" and "man". For example, the three indicators of the upper section are said to be located on both sides of the forehead (heaven), both sides of the cheek (earth) and in front of the ears (man) (*Suwen* 20; Unschuld and Tessenow 2011, p. 354). Altogether nine different pulse indicators are used to examine internal organs and various body parts (mouth, ears, eyes, etc.).

The "three sections and nine indicators", proving to be rather cumbersome, was simplified by the author of *Nanjing*, who reduced it to taking the pulse only at the *cunkou* (lit. inch-opening) area, that is, on the wrist over the radial artery (*Nanjing* 1). Zhang Zhongjing also tried to simplify the method of the "three sections and nine indicators" by reducing the sites of pulse-taking to three: at the *renyin* area on the neck, at the *cunkou* area on the wrist and at the *fuyang* area above the instep. Wang Shuhe developed his own pulse technique by dividing the *cunkou* area into *cun* (inch), *guan* (pass) and *chi* (cubit) sites, mirroring the spatial organization of specific organs. The upper position (*cun*) depicts the heart and lung, the middle (*guan*) represents the liver and spleen and the lower (*chi*) represents the kidney and the *mingmen* (gate of vital force). These *zang* organs are paired with their *fu* organs (*Maijing* 1) (Figure 3.3).

Wang Shuhe's pulse diagnostic method has proved to be particularly influential. Physicians learned to press their index, middle and ring fingers, with varying levels of pressure, on the patient's radial artery to feel their internal organs through the movement of the channels connecting them. Needless to say, it takes considerable time and practice to be able to correctly interpret the information gathered from pulse-taking and from the other three diagnostic methods. A good physician would have possessed the knowledge necessary to discern the true nature of the patient's disease and the appropriate therapy via these four approaches.

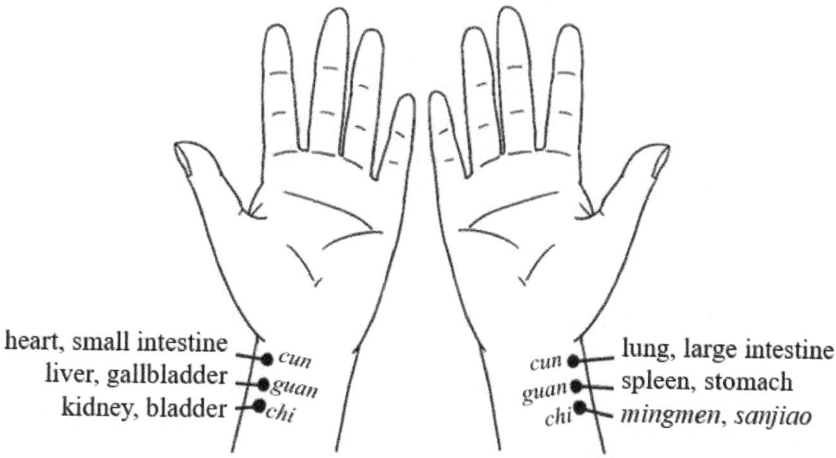

Figure 3.3. *Mapping of the zang–fu organs on the wrists, according to Wang Shuhe's Maijing*

3.7. Therapy

A wide variety of therapies were available during the Han period. Similarly, Qian's biographies of Bian Que and Chunyu Yi in *Shiji*, around the 1st century BCE, mention a physician of antiquity called Yu Fu, who, rather than curing illness with decoction, medicinal wine, stone needles, therapeutic exercise, massage or hot pack, instead conducted radical surgical operations (*Shiji* 105). From this anecdote, we can infer that the therapies listed by Sima Qian, apart from surgery, which would have been not only potentially dangerous but also contradictory to Confucian ideology, were employed by the Han physicians of his time.

Neijing mentioned at least seven types of therapies: *daoyin* (which often involves stretching and contracting the body and limbs, self-massage and controlled breathing) and moving *qi* (i.e. breath cultivation), massage, cauterization/moxibustion, hot packs, needling/acupuncture, burning therapy and decoction (*Lingshu* 42). As different environments gave rise to different diseases, various therapies were developed. *Neijing* discusses the origins of five different therapies based on the theory of the five agents: pointed stones therapy originated in the East, treatment with toxic drugs originated in the West, cauterization originated in the

North, nine needles therapy originated in the South and *daoyin* and massage originated in the central region (*Suwen* 12).

Lu and Needham (1980, pp. 1–2) suggest that the *Neijing* account of medical regionalism represents "a number of 'local culture', environing societies which brought various distinguishable traits into the eventual common Sinic stock". They postulated that "acupuncture would have been associated with the southeastern quasi-Indonesia aquatic element, while moxa would have come down to join it from the northern quasi-Tungusic nomadic element, and the pharmaceutical influence would have come from the western Szechuanese and quasi-Tibetan element" (ibid.).

Although excavated Han medical manuscripts describe diverse treatments, including household remedies, superficial surgery, demonic and spirit healing, *daoyin*, breath control and sexual cultivation, there is almost no mention of acupuncture, a technique that involves piercing the body with stones or needles to influence the *qi* in the body (Lo 2013). Acupuncture, which emerged out of a diverse and pluralistic medical landscape during the Han period, was a product of "*qi* and *yin-yang* practices, divination, numerology, minor surgery, and bloodletting, plus elements of spirit healing" (ibid., p. 47).

Jiu, translated as moxibustion or cauterization, is a form of heat therapy that is often performed by placing burning mugwort in the shape of a cone on strategic points of the body. Compared with acupuncture, moxibustion was the more accessible and widespread technique, and was recorded in the section of the Mawangdui medical manuscripts, where 11 channels were described (ibid.; Harper 1998).

The concept of medicine during the Han period can be placed in a broad context and is closely associated with the longevity and immortality practices of the *xian* (transcendence), which swept through the late 3rd century BCE elite society (Harper 1998, p. 50). Both Qinshi Huangdi and Emperor Wu of the Han (r. 140–87 BCE) were portrayed by Sima Qian as being obsessed with the search for drugs that would make them immortal. Medical texts in the "recipes and techniques" section of the bibliographic treatise in the Book of Han are listed under headings that include medical classics, canonical recipes, arts of the bedchamber and transcendents (Hanshu 30), the last category containing texts on *daoyin* exercises and massage.

Not all therapies would have been administered by physicians. Some were performed by individuals, in this case, members of the Han nobility, striving not only for good health but also for immortality. Such an understanding is also reflected in the classification of medicinal substances in the pharmacological writings of the Han and beyond.

3.8. Pharmacopoeia

In excavated medical manuscripts, there are numerous examples of the use of medicinal substances to treat ailments. Nearly 400 substances are recorded in Mawangdui manuscripts under the categories of plant, animal, mineral and prepared substances such as condiments and food (Harper 1998, p. 99). One of these manuscripts, *Wushier bingfang* (*Recipes for Fifty-two Ailments*), contains 283 recipes that use drugs and other therapies (ibid., p. 24). For example, one of the recipes for urine retention is to "take snails, twice seven in number and one small bunch of scallions. Boil them together in liquor and drink it" (Harper 1998, p. 256).

However, it was not until the 1st century CE that systematic relatively stylized writing on *materia medica* was undertaken, as evidenced in the *Shennong bencao jing* (Divine Farmer's Classic of *Materia Medica*). The term *bencao* (lit. roots and grass) refers both to knowledge about *materia medica* and to the genre producing such knowledge. It first appeared in the Book of Han (ca. 82 CE) when Emperor Ping of the Han (r. 1 BCE–6CE) issued an edict to recruit capable individuals with specialized knowledge, including those knowledgeable in *materia medica* (*Hanshu* 12). These *bencao* specialists were known as "*Materia medica* specialists awaiting edicts" (*bencao daizhao*). It is possible that *Shennong bencao jing* was compiled by such specialists working in the imperial court toward the end of the Western Han period (Liu 2021, p. 26).

The text contains 365 medicinal substances, corresponding to the number of days in a year, thus resonating with ideas of "correlative cosmology". These substances, such as plants, minerals, insects and animals, grains and foodstuff, were divided into three hierarchical rankings:

> There are 120 kinds of upper-class drugs. They are the lords which govern the nourishment of life. They correspond to Heaven and are not toxic. They are not harmful even if taken in large quantities or for a long period of time. Those who wish to make their body light, strengthen their *qi*, and prolong their lives, prevent aging, should have drugs from this upper class.
>
> There are 120 kinds of middle-class drugs. They are the ministers which govern the nourishment of human nature. They correspond to people. Some are without toxic; some possess toxic. One needs to gauge their proper use carefully. Those who wish to prevent illness and replenish bodily weakness should have drugs from this middle class.

> There are 125 kinds of lower-class drugs. They are the assistants and envoys which govern the curing of illnesses. They correspond to earth and are mostly toxic. One cannot take them for a long time. Those who wish to eliminate deviant *qi* caused by cold and heat, to counteract stagnation, and to cure illness, should have drugs from this lower class (*Shennong bencao jing*, preface; based on Tao Hongjing's *Bencao jingji zhu*).

The word *du*, rendered here as toxic, means "a double-edged sword". All lower class drugs, possessing *du,* have therapeutic benefits and are used to treat a variety of diseases. They would be better understood as "potent drugs" (Liu 2021).

In addition to classifying drugs in terms of whether or not they possess *du*, each drug is also characterized by one of the five flavors (sour, bitter, sweet, pungent, savory), which correspond to the five *zang* organs, and by one of the four "temperatures" (cold, hot, warm and cool), which correspond to the four seasons. In regard to combining different drugs for a particular recipe, the text advises that a harmonious and balanced prescription should have "one ruler, two ministers and five assistants" or "one ruler, three ministers and nine assistants" (*Shennong bencao jing*, preface).

The text also discusses the different forms in which drugs can be made and taken, such as pills, powder, boiled with water, soaked in wine and cooked into pastes.

The rich empirical knowledge found in *Shennong bencao jing* was foundational to the theory and practice of Chinese pharmacology. Its basic framework of categorizing and defining drugs has, in many ways, dictated all later *materia medica* writings.

3.9. Conclusion

Between the 3rd century BCE and the 3rd century CE, medicine and healing in China underwent extraordinary development and transformation, resonating with the profound political changes that accompanied the transition from the small feudal states of the late Warring States period to the Confucian–Legalist administration system of the Han empire. Ideas began to shift from attributing the cause of illness to the wrath of ancestors and demons to understanding the body as a microcosm of the cosmos, articulated in the language of *qi*, *yin–yang* and the five agents. Manuscripts and artifacts from the late Warring States and early imperial period, which have been excavated in recent decades, provide unique insight into this period of transition.

Physicians of the late Warring States, Qin and Han periods, who shared the sociopolitical and intellectual milieu of philosophers and technical experts, adopted and adapted philosophical and political ideas, which they further developed into the "medicine of systematic correspondence". The conceptualization of new medical theories, such as the "theory of viscera and their manifestations" and the "theory of channels and collaterals", was based on analogical reasoning rather than anatomical observations. The theory of the four seasons was closely connected to the *yangsheng* self-cultivation practices that formed an integral part of the new medicine of this period.

The Han dynasty was a crucial watershed for the formation of canonical works that employed the language of the new "medicine of systematic correspondence". The medical writings that later became "canonical" were compiled during the Han period. *Neijing*, the oldest received classic of medical theory, is a collection of heterogeneous medical writings by different authors, who often used the literary device of using an authoritative figure to advance their own ideas. As China became unified and centralized, so did medical theory and writing became more unified and systematic. Works such as *Nanjing* demonstrate a deliberate effort to consolidate various aspects of theoretical and practical knowledge into a more coherent and systematic doctrine. By the end of the Han, medical writings were produced on specific topics, such as the pulse, acupuncture and moxibustion, exogenous disease, and *materia medica*. All these texts laid a strong foundation for the theory and practice of Chinese medicine today.

The notion of medicine during the late Warring States and the Qin and Han periods encompassed much more than medical treatment by physicians. It was closely connected with the longevity and immortality practices of the *xian* cult, which emerged around the late 3rd century BCE. Self-cultivation practices, such as breathing exercises, *daoyin*, diet and sexual hygiene, constituted an integral part of medicine and healing during this period. Medicine was more than about curing disease; it was about preventing illness and about achieving immortality of the physical body, like the transcendent whose body was light enough to fly into the sky.

3.10. References

Pre-modern sources

Guanzi jinzhu jinyi 管子今註今譯 (Modern Annotated and Translated Master Guan). Guan Zhong 管仲 (c. 725–645 BCE). Li Mian 李勉 (717–788 CE) (ed.) (2013).

Hanshu 漢書 (Book of Han). Ban Gu 班固 (32–92 CE) *et al.* (1962). Zhonghua shuju, Beijing.

Huangdi bashier nanjing 黃帝八十一難經 (Yellow Emperor's Classic of Eighty-One Difficulties). 1st century CE (2007). Xueyuan chubanshe, Beijing.

Huangdi neijing lingshu 黃帝內經靈樞 (Inner Canon of the Yellow Emperor: Numinous Pivot). Wang Bing 王冰 (fl. 762.) (ed.) (1995). Zhongyi yanjiuyuan Ming edition repr. Zhongyi guji chubanshe, Beijing.

Huangdi neijing suwen 黃帝內經素問 (Inner Canon of the Yellow Emperor: Plain Questions). Wang Bing 王冰 (fl. 762.) (ed.) (1995). Zhongyi yanjiuyuan Ming edition repr. Zhongyi guji chubanshe, Beijing.

Lüshi chunqiu jiao shi 呂氏春秋校釋 (Collation and Annotation of the Master Lü's Spring and Autumn Annals). Lü Buwei 呂不韋 (292–235 BCE). Chen Qiyou 陳奇猷 (ed.) (1984). Xuelin chubanshe, Shanghai.

Maijing 脈經 (Pulse Classic). Wang Shuhe 王叔和 (ca. 180–270 CE) (1965). Taiwan shangwu yinshuguan, Taipei.

Maishu 脈書 (The Channel Book) 3rd century BCE. Jiangling Zhangjiashan Hanjian zhengli xiaozu 江陵張家山漢簡整理小組 (1989). Jiangling Zhangjiashan Hanjian Maishu shiwen 江陵張家山漢簡脈書釋文, *Wenwu*, 7, 72–74.

Mawangdui Hanmu Boshu 馬王堆漢墓帛書 (Silk Manuscripts from the Mawangdui Han Tombs). Mawangdui boshu zhengli xiaozu (1985). Wenwu chubanshe, Beijing.

Mengzi 孟子 (The Mencius). Mencius (ca. 372–289 BCE). In *Mengzi zhushu fujiao kanji shisan jing jiaoshu: fujiao kanji (Nanchang fuxue kanben)* 孟子注疏附挍勘記十三經注疏: 附校勘記 (南昌府學刊本). Ruan Yuan 阮元 (ed.) (1815/1976). Yiwen yinshu guan, Taipei.

Shanghan zabing lun 傷寒雜病論 (Treatise on Cold Damage and Miscellaneous Diseases). Zhang Zhongjing 張仲景 (150–219) (2014). Kexue jizhu chubanshe, Beijing.

Shen Nong bencao *jing* 神農本草經 (The Divine Farmer's Canon of *Materia Medica*), in Tao Hongjing (ed.) (456–526 CE) *Bencao jing jizhu* 本草經集注 (Collected Annotations on the 'Classic of *Materia Medica*'). Shang Zhijun 尚志鈞 and Shang Yuansheng 尚元勝 (eds) (1994). Renmin weisheng chubanshe, Beijing.

Shiji 史記 (Records of the Grand Historian). Sima Qian 司馬遷 (?145–ca. 86 BCE) (1962). Zhonghua shuju, Beijing.

Yinshu 引書 (The Book of Pulling). 3rd century BCE. Jiangling Zhangjiashan Hanjian zhengli xiaozu 江陵張家山漢簡整理小組 (1990) 'Jiangling Zhangjiashan hanjian yinshu shiwen 江陵張家山漢簡引書釋文', *Wenwu*, 10, 82–86.

Zhenjiu jiayi jing 針灸甲乙經 (AB Canon of Acupuncture and Moxibustion). Huangfu Mi 皇甫謐 ((215–282 CE). Xu Guoqian 徐國仟 and Zhang Canjia 張燦玾 (eds) (1996). Renmin weisheng chubanshe, Beijing.

Modern sources

Graham, A.C. (1989). *Disputers of the Tao: Philosophical Argument in Ancient China*. Open Court, La Salle, IL.

Harper, D. (1998). *Early Chinese Medical Literature: The Mawangdui Medical Manuscripts*. Kegan Paul International, London.

Harper, D. (1999). Warring States natural philosophy and occult thought. In *The Cambridge History of Ancient China: From the Origins of Civilization to 221*, Loewe, M. and Shaughnessy, E.L. (eds). Cambridge University Press, Cambridge.

Hinrichs, T.J. and Barnes, L.L. (eds) (2013). *Chinese Medicine and Healing: An Illustrated History*. The Belknap Press of Harvard University Press, Cambridge, MA/London.

Hsu, E. (2010). *Pulse Diagnosis in Early Chinese Medicine: The Telling Touch*. Cambridge University Press, Cambridge.

Keegan, D.J. (1988). *The 'Huang-Ti Nei-Ching': The Structure of the Compilation; The Significance of the Structure*. University of California Press, Berkeley.

Kuriyama, S. (1999). *The Expressiveness of the Body and the Divergence of Greek and Chinese Medicine*. Zone Books, New York.

Li, Z. and Kong, Y. C. (2010). *Neijing zhiyao yigu* 內經知要譯詁 (*Huangdi neijing*: A Synopsis with Commentaries). Chinese University Press, Hong Kong.

Liao, Y. 廖育群 (1988). Jinben Huangdi neijing yanjiu 今本《 黄帝内经》 研究. *Ziran kexueshi yanjou*, 4, 367–374.

Liao, Y. 廖育群 (2005). Lianhan yixueshi de chonggou 两汉医学史的重构. *Kexue wenhua pinglun*, 4, 46–64.

Liu, Y. (2021). *Healing with Poisons: Potent Medicines in Medieval China*. University of Washington Press, Seattle, WA.

Lo, V. (1998). The influence of Yangsheng culture on early Chinese medical theory. PhD Thesis, The University of London, London.

Lo, V. (2013). The Han period. In *Chinese Medicine and Healing: An Illustrated History*, Hinrichs T.J. and Barnes, L.L. (eds). The Belknap Press of Harvard University Press, Cambridge, MA/London.

Lo, V. (2014). *How to do the Gibbon Walk: A Translation of the Pulling Book (ca. 186 BCE).* Needham Research Institute, Cambridge.

Lo, V. (2024). Looms of life: Weaving a new medical imaginary. *Asian Medicine*, 18(1/2), 148–166.

Lo, V., Stanley-Baker, M., Yang, D. (eds) (2022). *Routledge Handbook of Chinese Medicine.* Routledge, London.

Lloyd, G.E.R. and Sivin, N. (2002) *The Way and the Word: Science and Medicine in Early China and Greece*. Yale University Press, New Haven, CT.

Lu, G. and Needham, J. (1980) *Celestial Lancets: A History and Rational of Acupuncture and Moxa*. Cambridge University Press, Cambridge.

Sivin, N. (1987). *Traditional Medicine in Contemporary China: A Partial Translation of Revised Outline of Chinese Medicine (1972) with an Introductory Study on Change in Present-Day and Early Medicine*. Center for Chinese Studies, University of Michigan, Ann Arbor, MI.

Sivin, N. (1993). Huang ti nei ching. In *Early Chinese Texts: A Bibliographical Guide*. Loewe, M. (ed.). The Society for the Study of Early China and The Institute of East Asian Studies, University of California, Berkeley, CA.

Sivin, N. (1995). State, cosmos, and body in the last three centuries B.C. *Harvard Journal of Asiatic Studies*, 55(1), 5–37.

Sivin, N. (2000). Book review: Christoph Harbsmeier, science and civilisation in China, vol. 7, the social background, part 1: Language and logic in traditional China. *East Asian Science Technology and Medicine*, 17, 121–133.

Stanley-Baker, M. (2022). Qi 氣: A means for cohering natural knowledge. In *Routledge Handbook of Chinese Medicine*, Lo, V., Stanley-Baker, M., Yang, D. (eds). Routledge, London.

Unschuld, P.U. (1985) *Medicine in China: A History of Ideas*. University of California Press, Berkeley, CA.

Unschuld, P.U. (1986) *Nan-Ching: The Classic of Difficult Issues: With Commentaries by Chinese and Japanese Authors from the Third through the Twentieth Century*. University of California Press, Berkeley, CA.

Unschuld, P.U. (1998). *Chinese Medicine*, Wiseman, N. (trans.). Paradigm Publications, Brookline, MA.

Unschuld, P.U. (2003). *Huang Di Nei Jing Su Wen: Nature, Knowledge, Imagery in an Ancient Chinese Medical Text, with an Appendix, the Doctrine of the Five Periods and Six Qi in the Huang Di Nei Jing Su Wen*. University of California Press, Berkeley, CA.

Unschuld, P.U. (2016). *Huang Di Nei Jing Ling Shu*. University of California Press, Berkeley, CA.

Unschuld, P.U. and Tessenow, H. (2011). *Huang Di Nei Jing Su Wen: An Annotated Translation of Huang Di's Inner Classic – Basic Questions*, 2 volumes. University of California Press, Berkeley, CA.

Wilms, S. (2016). *The Divine Farmer's Classic of Materia Medica: A Translation of the Shen Nong Bencao Jing*. Happy Goat Productions, Corbett, OR.

Yamada, K. 山田慶兒 (1979). The formation of the Huang-ti Nei-ching. *Acta Asiatica*, 36, 67–89.

Yang, D. (2018). *Prescribing 'Guiding and Pulling': The institutionalisation of therapeutic exercise in Sui China (581–618 CE)*. PhD Thesis, University College London, London.

4

India's Early Medical History: Buddhism and the Beginnings of Ayurveda

Kenneth G. ZYSK
University of Copenhagen, Denmark

4.1. Introduction

In the centuries just prior to and after the beginning of the Common Era, an important transition and paradigm shift took place in the history of Indian medicine, where a new mode of medical thinking superseded the old. We want to understand the reasons and mechanisms for that change, and the role Buddhism played in it.

For this, we will require a brief orientation on the basic terminology and sources mentioned in this chapter. The word *āyurveda* (hereafter Ayurveda), meaning the knowledge of human life, is the Sanskrit term that designates a new type of medical thinking first codified by learned doctors in the Sanskrit medical compilations of Caraka and Suśruta around the beginning of the Common Era. The modern tradition of Ayurveda traces its medical roots back to these early authors. Buddhism is a religious movement, traditionally attributed to a wise prince by the name Siddhārtha Gautama, who was the founder of a philosophy of life that emphasized the avoidance of extremes, known as the Middle Way, in the northern part of the Indian subcontinent around the 5th century BCE. His teachings inspired many followers, who subsequently organized themselves into a coenobitic lifestyle governed by a

History of Life Sciences and Medicine 1,
coordinated by Lorenzo PERILLI. © ISTE Ltd 2026.

Code of Behavior or Monastic Code and informed by the founder's teachings, composed in the Pāli language (a dialect of Sanskrit), probably shortly after the time of the Buddha and finally written down in Sri Lanka in the 1st century BCE. Parts of these books record the medical doctrines and practices known to early Buddhists. Before delving into Buddhist medicine and its importance in the development of Ayurveda, it is prudent to have idea of the four phases of Indian medical development.

4.2. Four phases of Indian medicine

These are the four phases of Indian medical history.

4.2.1. *I. Vedic medicine (c. 1200–500 BCE)*

Symbolic medicine with emphasis on exorcism. Information about medicine during this period derives from numerous curative incantations and references to healing rituals found in the ancient Sanskrit treatises of the *Atharvaveda* and, to a much lesser extent, the *Ṛgveda*, and carried into the later Vedic ritual literature of the *Sūtras*.

4.2.2. *II. "Classical" Ayurveda (c. 200 BCE–10th century CE)*

The advent of the first Sanskrit medical treatises, the *Caraka Saṃhitā* and *Suśruta Saṃhitā*. This period includes all subsequent medical treatises dating from before the Muslim invasions of India at the beginning of the 11th century. The later works tend to closely follow the systematic presentation of medicine in earlier classical compilations.

Principal works

A) *Caraka Saṃhitā* (c. 100 BCE–200 CE).

B) *Suśruta Saṃhitā* (c. 400 CE).

C) Syntheses:

1) *Aṣṭāṅgasaṃgraha* (Vāgbhaṭa) (c. 7th century CE).

2) *Aṣṭāṅgahṛdayasaṃhitā* (Vāgbhaṭa) (c. 7th century CE).

4.2.3. *III. "Syncretic" Indian medicine (11th century–20th century)*

There were clear influences on the classical paradigm from Islamic or Unani and other non-classical medical traditions, as witnessed in Śārṅgadhara's *Śārṅgadhara Saṃhitā,* dating from the 14th century, and Bhāvamiśra's 16th century *Bhāvaprakāśa.*

4.2.4. *IV. "New Age Ayurveda" (20th century–21st century)*

Classical Ayurveda was imported to the West and re-imported into India, with influences from complementary and alternative medicine.

4.3. Vedic medicine

The earliest literary evidence of medicine on the Indian subcontinent can be traced with certainty to the Vedic Period (c. 1500–500 BCE), exemplified by treatises composed in the earliest form of the Indo-European language of Sanskrit. In the face of a language that has not been deciphered, archaeological remains from the earlier period of the Indus civilization provide inconclusive information.

The Vedic medical lore comes primarily from the *Atharvaveda*, which is devoted to religious matters, but also contains information pertaining to medicine. Best estimates put its compilation no later than the 9th century BCE; however, it probably looks back to an earlier period.

The predominant type of healing described in this work is what is best described as symbolic medicine. Disease was not conceived to be a physiological disorder, but was understood to result from malevolent forces that invaded and took over the body from the outside and caused morbid physical states. Reminiscent of exorcism known the world over, the removal of such malign entities involved elaborate rituals aimed at coaxing the demon out of the body, using special amulets usually composed of parts of animals and plants similar in color and form to that of the perceived demon of the disease, manifested through different bodily symptoms. With these apotropaic devices and techniques, the demon afflicting the body was then removed.

4.4. Ayurveda

Sometime around the beginning of the Common Era, a dramatic shift in medical thinking began. No longer were magic and religion the predominant means for healing, although they never disappeared completely from Indian medicine, cropping up again in the treatment of children's afflictions in the early compilations of Ayurveda. The ideology and rituals of Vedic medicine were superseded by a mode of thinking that understood the world in terms of a micro-macrocosm interaction of basic elements, which is a cornerstone of Ayurveda to the present day.

For the first time, medical knowledge was collected and formulated into specialized books, the earliest of which included two Sanskrit medical corpora bearing the name of their earliest compilers, Caraka and Suśruta. These two savants of medicine assembled the existing medical knowledge into coherent treatises divided into chapters dealing with specific aspects of pharmaceutics and therapeutics. The latter had an additional emphasis on healing by surgical methods. Common to both treatises is the etiology based on pathogenic agents (*doṣa*s).

The human body was conceived to be derived from five basic elements: earth, water, fire, air and space or ether, which combined to produce the seven root bodily elements (*dhātu*s) from digested food, plasma, blood, flesh, fat, bone, marrow and semen. A healthy person was considered to have these seven root bodily elements functioning in a constructive manner, leading to health and well-being.

As in Hippocratic medicine, digestion played a crucial role in Ayurveda. Digestion broke down and transformed food with a type of cooking fire (*agni*) located in the region of the stomach, from which two kinds of substances were produced: root bodily elements and waste products, which included the three pathogenic agents whose qualities include wind (*vāyu, vāta*), bile (*pitta*) and phlegm (*kapha, śleṣman*), which sometimes acted in unison or in pairs, along with urine, sweat and feces.

In a normal healthy state, the three pathogenic agents remained stable in their original seats in the body, a condition defined as a person's natural state (*prakṛti*). The change from the natural to the unnatural state (*vikṛti*) could be caused by several factors that were climatic, organic, external, or, less commonly, demonic or supernatural in origin. These factors cause one or more pathogenic agents to become agitated and riled, whence it left its original seat only to manifest elsewhere in the body and cause morbidity or disease.

The nosology of the three pathogenic agents was applied to treatment procedures and was the principal means for diagnosing morbid or adverse changes in the patient's daily bodily regimen.

The same ideology applied to the natural world led to an understanding of the basic elements in the macrocosmic world of plants and animals. With the aim of balance, represented in the patient's natural state, the physician (*vaidya*) had the primary responsibility to identify which of the pathogenic agents became riled and dislodged, to remove one or more of them through purgation therapies, and finally to calm and restore the body to its natural state. To do this, he used a combination of purgation and bodily cleansing, followed by medicines and diet, which in the beginning were organic (both meat and plants), but eventually included inorganic or alchemical substances. Ayurvedic therapy involved both intervention and the use of pharmaceuticals.

The predominant form of early ayurvedic intervention involved the application of five modes of body purification (*pañcakarman*) in preparation for the use of pharmaceutics. The five procedures involved head purgation, vomiting, dry and wet enemas, and bloodletting.

Both early medical compilations contained detailed medical formulae and the process by which the different medicines were to be prepared. The traditional Sanskrit pharmacopoeia was extensive and grew even larger and more detailed over time, taking in regional variations from over the entire Indian subcontinent.

The list included both organic and inorganic substances, which came from different places within India and neighboring lands. These were eventually assembled and described, along with different synonyms and the diseases they cured, in dictionaries of materia medica called *Nighaṇṭu*s or Medical Glossaries.

The transmission of medical knowledge continued in Sanskrit with the subsequent compilations of Vāgbhaṭa, who prepared two compendia of Ayurveda in the late 7th century, some 500–700 years after the original medical collections. One in verse provided another formulation of Caraka's collection with the addition of new material. It is called "The Collection of the Heart of the Eight Limbs (of Ayurveda)" (*Aṣṭāṅgahṛdayasaṃhitā*). The other in both verse and prose was essentially a version of the first, but also included material from Suśruta's compilation. It is called "The Collection of the Eight Limbs (of Ayurveda)" (*Aṣṭāṅgasaṃgraha*). To this day, the ayurvedic tradition knows the three authors Caraka, Suśruta and Vāgbhaṭa as the "Great Triad" of Ayurveda.

4.5. Historic transition

From the perspective of medical history, it is immediately apparent that a significant change or paradigm shift had taken place in medical theory and practice, where Vedic medicine was superseded by the systematic presentation of a comprehensive form of pharmaceutics and therapeutics in Ayurveda. The change involved the sidelining of Vedic medicine and the prominence of Ayurveda. The obvious questions to ask are how such a change in medical thinking took place and where the new ideas came from. There is little to inform us in the Sanskrit literature from the later Veda period to early texts of early Ayurveda (c. 500–200 BCE). What does exist in this Brahmanic literature reflects the viewpoint of priests, who considered physicians to occupy a low social status because of their contact with polluted people and impure body substances. Rather than relying on the Sanskrit sources, we will have to look in another direction, toward thinking that is not dominated by Brahman priests. This alternate mode of thought is found in the Pāli Canon of the early Buddhists.

4.6. Buddhism's contributions to Early Indian medicine

Buddhism played a crucial role in the development of this new medicine of Ayurveda. Coming into existence as a religious movement between approximately 500 and 200 BCE, it provides a glimpse of medicine just prior to and during the earliest compilations of Ayurveda.

The Buddhist literature of the Pāli Canon repeatedly refers to medicine and healing. For the medical historian, a significant collection of information is assembled and preserved in the first book dealing with the Monastic Code, called the "Big Chapter" (*Mahāvagga*), which, pertaining to a regimen of life for monks and nuns, is among the earliest parts of the Canon. It reflects a time when the assemblies of coenobites, called Saṅgha, were being founded and the rules to govern their monastic life were being set.

One chapter of this book is called "Medicines" (*bhesajja*). It contains a pharmacopeia followed by therapeutics recorded on a uniquely case-by-case basis, presented in a question-and-answer format in which the Buddha assumed the role of a doctor. A typical case involved a sick monk who approached the Buddha for treatment. The Buddha diagnosed the problem and prescribed the appropriate medicines and mode of treatment. The recording of this information represents India's first medical textbook.

Although brief, the content of these cases resembles the medicine described in the early ayurvedic Sanskrit medical treatises of a slightly later period. The similarity between them points to their common origin and provides valuable clues for understanding the beginnings and process of transmission of a new form of medical thinking. For the first time, Buddhist documents present medical knowledge in a coherent and systematic manner on the basis of familiarity with a nosology involving pathogenic agents (wind, bile and phlegm). Moreover, Buddhist medicine is a complete healing system that includes both pharmaceutics and therapeutics.

4.6.1. *Buddhist pharmaceutics*

The earliest form of a pharmacy was a kind of storehouse of foods and medicines, described in the above-mentioned chapter of the Monastic Code, devoted to medicines. Since the rules forbade coenobites from eating after noon, the consumption of foodstuffs after that time had to have a special dispensation, which came in the form of their designation as medicines (Pāli *bhesajja*), which allowed certain foodstuffs to be consumed around the clock. The use of materia medica, which began with five basic medicines, eventually increased over time to include plant, animal and mineral products.

The five basic medicines consisted of ghee or clarified butter (*ghṛta*), honey (*madhu*), fresh butter (*navanīta*), oil (*taila*) and molasses (*phāṇita*). Medicines divided into fats, roots, extracts, leaves, fruits, resins and salts were added to these.

Their method of classification into different substances is reflected in the system of drugs (*dravya*) found in the two early ayurvedic works and continues into modern Ayurveda. Buddhist sources permit us to understand how the process of organization and classification of medicine began among the earliest Buddhist ascetics.

4.6.2. *Buddhist therapeutics*

Following the materia medica, medical therapeutics are offered on the same case-by-case basis. The prescribed remedies presumed a nosology that involved a fourfold system of pathogenic agents (Pāli *dosa*) mentioned in another part of the Canon devoted to the teachings of the Buddha.

When asked by the ascetic Sīvaka, if past actions (*kamma,* Skt. *karman*) were the sole causes of human suffering as some say, the Buddha responded by rejecting this point of view in favor of his explanation that humankind's suffering is eightfold:

> Bile, phlegm, wind, and their simultaneous corruption, as well as season, imbalance, external agency, and the ripening of past actions as the eight.[1]

This medical teaching of the eight causes of disease contains the earliest formulation of the ayurvedic nosology of the pathogenic agents. Here, there are three elements, wind, bile and phlegm, plus a fourth, their simultaneous corruption. A fourfold formulation of the elements also occurs in the early ayurvedic text of the

---

1 Zysk (2021b, p. 39).

Suśruta Saṃhitā, where blood (*śoṇita*) replaces simultaneous corruption. The mention of blood as a pathogenic element immediately gives rise to the question of the possible influence of early Greek medicine around the time of the Hippocratic Corpus. In the end, however, the formulation with only three pathogenic agents has prevailed throughout Ayurveda to date. The threefold formula is modeled on the enumeration of the three basic qualities (*guṇa*) of Indian philosophy, which states that all existence is made up of three distinct qualities: creativity (*sattva*), passion (*rajas*) and dullness (*tamas*).[2]

Early Buddhists had access to a form of medicinal treatment in which both method and content rather closely resembled the nosology and therapeutics of the early ayurvedic treatises of Caraka and Suśruta.

4.6.3. *The case-by-case methodology of early Buddhist medicine*

The case-by-case presentation of medicine illustrates an overall practical approach to Buddhist healing. This is exemplified in two sections of the early Buddhist Monastic Code. One offers Buddha's cures, and the other offers those of the famous lay physician, Jīvaka, The Pediatrician, whose healing methods are retold in Buddhist literature from South to South-East and East Asia. The two provide a clear view of Indian medicine a few centuries before the earliest parts of the Sanskrit medical books of early Ayurveda.

The Buddhists, whose central philosophy emphasized a Middle Way between pleasure and pain, and whose religious practices focused on asceticism and meditation, were the main vehicles for the transmission of an entirely new method of identifying and removing human suffering from the human body. From the perspective of historical epidemiology, these documents offer indications of the types of human afflictions that wandering mendicants especially suffered from in ancient India.

4.7. The Buddha's case histories

The Buddhist disciplinary codes record eighteen different kinds of afflictions, diseases and ailments, for which a treatment was dispensed to Buddhist monks by the Buddha. The procedure of each case was the same: a certain monk, suffering from a morbid bodily condition with a certain set of symptoms, came to the Buddha for treatment. After considering the matter, the Buddha then sanctioned a particular

2 Zysk (2021a).

course of treatment, which included medicines and therapeutics. Once permitted, the disease and its cure were noted and preserved as part of the Monastic Code, which then allowed the same treatment to be used repeatedly. In this way, the ordinances of the world's earliest monastic order became, at the same time, the world's first handbook of medical therapeutics and served as a model for the ayurvedic handbooks of materia medica, which were essentially early versions of an Indian medical vade-mecum.

The cases are grouped randomly for the most part, beginning with large sores and ending with rectal fistulae. However, some illnesses occur in groups, suggesting that parts may have been extracted from existing medical collections that are now lost. These included the set of three wind diseases: affliction of wind, wind in the limbs and wind in the joints, plus one out of sequence: wind in the abdomen; the group of three external afflictions: split-open feet, swellings and two types of poison (snakebite and artificial poisons); and two types of skin conditions: morbid pallor (jaundice?) and corruption of the skin. Diseases caused by pathogenic agents are encountered again when the body is filled with them and when there is out-of-sequence wind in the abdomen.[3]

The predominance of diseases caused by wind points to familiarity with a formulation of the pathogenic agents that include wind, and the general agreement of theory and practice in both the Buddhist scriptures and Sanskrit medical textbooks clearly points to a common source for both.

4.7.1. *An example from Buddha's cases*

> A certain monk suffered from a condition in which his body was filled with the pathogenic agents. He consulted the Buddha, who prescribed as treatment the drinking of a purgative, followed by a diet of clarified sour rice gruel, natural (i.e., non-oily) soup, both natural and unnatural (i.e. non-oily and oily) soup, and finally a meat broth.[4]

An examination of the chapters on emetics and purgatives in the early Sanskrit medical textbooks of Caraka and Suśruta indicates that both forms of evacuation were used interchangeably for various conditions involving the accumulation of the riled pathogenic agents in the body. The technique very closely follows the method

3 All of the disorders are listed in the appendix in the order they occur in the text.
4 Zysk (2021b, p. 148).

prescribed in the therapy that uses the "five actions" of bodily cleansing. Before the purgation, the patient would eat a type of gruel. He would next be massaged with oil and then be made to sweat by the use of a kind of steam bath in which coals are used to heat a bed of wet leaves to provide moisture. He was then purged by one or several of the five purgative actions and fed a meat broth to help him regain strength.

According to early medical treatises, a medium of soup was commonly used to administer both purgatives and emetics, and after purgation, the patient consumed either a vegetarian soup of green grams or a meat broth prepared from jungle animals.

It is clear from this comparison that both the Buddhist and ayurvedic techniques for eliminating and pacifying riled pathogenic agents relied on the same process. Importantly, even though Buddhists were commonly thought to have been vegetarians, in the case of medical treatment, they were not averse to consuming meat products, such as meat soups.

4.7.2. *Jīvaka's case histories*

Another set of six case-based treatments, found in the Monastic Code of the Buddhist texts in Pāli, is attributed to the physician Jīvaka, The Pediatrician. What we know about this phenomenal healer comes largely from stories and legends about him transmitted through all of Buddhist literature in almost every Asian language. By all accounts, he was a well-known doctor with many miraculous cures to his credit. One of the earliest and most extensive accounts is found in the Monastic Code of an early Buddhist sect from the northwestern part of the Indian subcontinent around the beginning of the Common Era. It describes Jīvaka's medical studies at Taxila, the ancient Indo-Greek town located in modern Pakistan and known as a principal seat of education and the arts, examples of which clearly reflect Hellenistic influences. In an early version of the story preserved in the Pāli language, Jīvaka is said to have cured many important people, including kings and the Buddha. The accounts of his treatments are nearly the same as those found in the early Sanskrit medical treatises, pointing again to a common source for both Jīvaka's medicine and classical Ayurveda.

4.7.3. *An example of Jīvaka's cases*

As with the Buddha's cases, Jīvaka's medicine involved familiarity with the nosology based on the pathogenic agents and closely resembled the early ayurvedic medicine detailed in the compendia of Caraka and Suśruta. His treatment of the Buddha is a case in point:

> The Buddha suffered from a condition in which his body was filled with the pathogenic agents. Jīvaka attended to him and administered a mild purgative, which involved first oiling the body for a few days, followed by the inhalation of the fragrance of three individual handfuls of lotus-blossoms mixed with various medicines, and finally a bath, resulting in thirty purges. He was then fed only soup until his body returned to a normal condition.[5]

In the general course of administering a purgative, as already mentioned, early ayurvedic treatises prescribed that a patient should be massaged with oil and made to sweat; after that he was to be fed light food, hot water and sour fruits on the day before purgation. The pre-purgative procedure aimed to loosen the pathogenic agents from their abnormal seats so that they could easily be purged from the body. After the administration of the purgative, the patient was to eat only light, lukewarm and thin gruel until the bodily constitution returned to its normal condition, and the pathogenic agents were returned to their natural state.

In the case of a patient with a delicate bodily constitution, such as that of an ascetic such as the Buddha whose diet was restricted to one (often light) meal a day, the medical texts prescribed a special purgative. The formula called for the use of the powder of the seeds of two plants, the emetic nut and the ribbed luffa, repeatedly soaked in a decoction of emetic drugs. This powder was then sprinkled over large lotus flowers. When given to the patient, they were requested to inhale the powder, causing them to purge by vomiting.

It is remarkable how close this treatment found in the medical treatises corresponds to that which Jīvaka administered to the Buddha for the same condition.

Jīvaka also performed extraordinary cures, such as alleviating a headache by means of a type of trepanation and a twisted bowel by a type of laparotomy. These cures, along with his successful treatment of the Buddha and two kings, were favorite stories told by Buddhists, wherever they wandered. These episodes of the doctor Jīvaka, along with their accounts of wonderful healings, obviously made Buddhists popular among the people with whom they came in contact with, and thereby greatly facilitated the rapid spread of Buddhism through Asia.

5 Zysk (2021b, p.172).

In the earliest version of the story, Jīvaka performed six cures; however, in the later Sanskrit version, the number increased to thirteen. They included the same type of trepanation and laparotomy, and Buddha's cure; however, Jīvaka also healed a blind man and a man whose eyeball came out of its socket, and removed incurable tumors from a householder and a king. He performed six cures in the story recorded in China, which included most of those found in the earliest Indian version. However, in the case of a boy who fell while playing hard and died, he diagnosed the cause of the problem as an injured liver with the help of a branch from a magic tree (a piece of uranium ore?) that permitted him to see inside the boy's body. He was cured by surgery and the application of three kinds of supernatural oils. The boy recovered completely after three days.

The early Buddhist texts provide the first real evidence of the medical system in India. The healing beliefs and practices of ancient India, represented in the Vedas, were superseded by an entirely new form of medical thinking first recorded in Buddhist sources from several centuries before the Common Era. It was this new medical system that became codified in the early medical treatises of Ayurveda.

4.8. Medicine in Later Buddhism

Because of medicine's important role in carrying out a central Buddhist doctrine to eliminate suffering, it quickly became part of Buddhist institutions, as they developed in India and elsewhere after the time of the Buddha. For many years, Buddhist medicine and Ayurveda developed alongside each other in two different religious settings. The latter eventually became distinguished from the former, when the medical texts began to assume a distinctive Hindu religious veneer and were transmitted via the Sanskrit language.

4.8.1. *Buddhist medical institutions*

As Buddhism grew and spread, first throughout India and then to southeast and central Asia, Tibet and China, medicine went with it, since it had already been codified as a part of Buddhist canonical scripture. It played an essential role in the spread of the religious system throughout Asia in much the same way that apothecaries and infirmaries aided in the missionary work of medieval Christianity.

In the middle of the 3rd century BCE, a virtuous king named Aśoka, who closely followed Buddha's teachings, provided medical treatment for both humans and animals throughout India and imported medicinal herbs, roots and fruits, where local remedies could not be found. By the 5th century CE, the Chinese Buddhist pilgrim

Faxian records that in the city of Pāṭaliputra (modern Patna in the north Indian state of Bihar), houses were established for dispensing charity and medicines to the poor, destitute and sick. Finally, in the early part of the 7th century, the Chinese Buddhist pilgrim Xuanzang reported that medicine was a significant part of the educational curriculum at the large Buddhist monastery at Nālandā also located in modern Bihar. The same curriculum became the core of education in the Buddhist monasteries of Tibet in the 14th century, when the Tibetan Buddhist historian Buston reported that the principal medicinal treatise taught at the monasteries was Vāgbhaṭa's *Collection of the Essence of the Eight Limbs (of Ayurveda)*, which, as previously mentioned, was a major second-level compilation, after Caraka and Suśruta, from around the 7th century CE.

The Buddhists, therefore, were both directly and indirectly responsible for a system of public health and for the establishment of medical education in ancient India, as well as other parts of Asia influenced by Buddhism.

4.9. Buddhist medical texts

The earliest Buddhist medicine recorded in the Monastic Code of the Pāli Canon was in the form of case histories. The medicine focused on practice rather than theory and provided coenobites with a type of medical manual for the treatment of their common ailments. This form of medical handbook developed within the Buddhist tradition and is a major Buddhist literary legacy to the Indian medical tradition.

When the Buddhist monastery became a permanent place, medicine became institutionalized. Infirmaries, sometimes called "health houses," were established to care for the sick and ailing, and specific medical texts were composed to teach medicine at larger monasteries in India.

Among the literary remains from Buddhist sites in Central Asia, we find medical manuals written in Sanskrit, Sanskrit mixed with Prakrit and Sanskrit with vernacular translations. Being essentially devoid of medical theory, focusing on therapy, they represented the continuation of the early model of practical medicine in the form of symptomatic treatment already established in the Monastic Code. However, they are considerably more extensive, covering a wide range of diseases with a full elaboration of the cure. Such treatises date from the 6th century CE. One of the earliest is the chapter on the [*Formulas*] *Derived from Fresh Butter* of the Bower Manuscript, preserved in Sanskrit. It comes from Central Asia and points to the Buddhist preservation of medical recipes in a form resembling the leech books of medieval England. Others included Nāgārjuna's *Hundred Formulas*, found in both Sanskrit and Tibetan versions; *Jīvaka's Medical Book*, found in Khotanese and corrupt Sanskrit; and Ravigupta's *Essence of Successful Medicine*, found in Sanskrit, Tibetan and Khotanese, and which includes a more detailed diagnosis by means of pathogenic agents.

In addition, there are fragments of medical manuals that have a Sanskrit text followed by a translation in the central Asian language of Tokharian. Some of these fragments appear to be translated from Nāgārjuna's *Hundred Formulas*.

The style of all these texts is basically the same. A recipe is given, which lists all the ingredients, usually in order of importance, and the precise way in which they should be combined and administered. A list of diseases for which the formula is efficacious follows the recipe. The same style occurs in the later ayurvedic books of materia medica, called Nighaṇṭus, which are used by many current ayurvedic doctors. The most well-known of these is Bhāvamiśra's 14th century *The Explanations of Bhāva*.

In addition to these practical medical handbooks, the Tibetan Buddhist Canon contains Tibetan translations of Vāgbhaṭa's *Collections of the Essence of the Eight Limbs [of Ayurveda]* and Nāgārjuna's *Hundred Formulas*. In Tibet and Mongolia, the principal medical treatise, *The Four Tantras*, is a composite work incorporating elements of Chinese medicine within the framework of Indian medicine. In the Chinese Buddhist Canon, numerous medical aphoristic texts are found, two of which focus on pediatrics. There is *The Sūtra Spoken by Rāvaṇa on the Curing of Children's Diseases*, which is a 10th-century Chinese translation of the Sanskrit treatise *Rāvaṇakumāratantra*, and *The Sūtra of Gynecology Taught by the Sage Kāśyapa*, which was also translated in the 10th century.

Beginning with the practical medical case histories contained in the Monastic Code of the Pāli Canon, a rich tradition of medical textual transmission and literary composition evolved in Buddhist monasteries. This led to the development of medical literature aimed at the practicing doctor and gave rise to the principal medical textbooks that offered formal instruction in major monasteries.

4.10. Discussion and conclusions

We have seen how the religious movement of Buddhism played a decisive role in Indian medical history. The early doctrines centering on the rules for living an ascetic's life in a cloistered environment included the healing and care of the sick. A formalized system for carrying out these duties involved the essentials of any medical system: pharmacopeia and therapeutics. I will, therefore, conclude this chapter with a deeper exploration of these two aspects of early Indian medicine.

4.10.1. *Pharmacopeia*

It is clear from the lists of medicines present in the early Buddhist literature that they belong to a collection of foodstuffs. Since Buddhists could only accept items from begging, foodstuffs included, among other things, medicines. These medicaments were eventually collected, classified, recorded and transmitted by wandering mendicants who acquired them from the people whom they met on their travels. When transient monks began to be more sedentary and cloistered in early monastic institutions, these medicines were codified and incorporated into the early Monastic Code.

One of the Buddhist's most important contributions to early Indian medicine was the development of a system for the classification of medicine using enumeration. The use of numbers to help organize and understand the world is very old in Indian thought, which reaches a high-water mark among Buddhist intellectuals. The cornerstone of Buddhist medicine, then, was its catalog of medicines, to which a method of therapeutics based on a nosology of a fixed number of pathogenic agents was applied.

4.10.2. *Theory*

The search for the theoretical and practical overlay to the medical formulae poses another set of obstacles, since such a system of disease causation did not exist in India prior to the Buddhists, except for the method of symbolic associations underpinning the medical charms and incantations of the Veda. The doctrine of the four pathogenic agents and the case-by-case organizational method of Buddhist medicine are two important elements that distinguish Ayurveda from the esoteric rituals of Vedic medicine.

4.10.3. *Nosology based on pathogenic agents*

Despite the threefold nosology (wind, bile and phlegm) that became the ayurvedic norm transmitted over time from Caraka's compilation, another old formulation involved four pathogenic agents already present in early Buddhist literature in the three-plus-one formulation: wind, bile and phlegm, plus a fourth, which in Buddhist medicine was the simultaneous corruption of the previous three, and which in Suśruta's Ayurveda took the form of blood. Both the fourfold nosology and the occurrence of blood as a pathogenic agent were unique in Indian medical thinking at the time, and point to ideas common to early Hippocratic and Greek medical nosology, which included, in most cases, four principal elements, later humors, that varied in different formulations but always involved three components: blood, bile and phlegm. By the time of Galen in the 2nd century CE, the standard humors were blood, phlegm, and black and yellow bile. The similarities between Greek and both Buddhist and Suśruta's nosology should not be underestimated. Although direct textual transmission is wanting, the overall congruence between the two theories and their modes of explanation point to some kind of exchange of medical ideas, probably in the northwest of the Indian subcontinent in the centuries bracketing the Common Era. It is, therefore, likely that the theoretical overlay on the Indian materia medica resulted from both indigenous thinking and the adaptation of non-indigenous medical notions.[6]

4.11. Summary

The particular method of recording the doctor's diagnosis and treatment by means of medical case histories is unique in Indian medical history and points to actual medicine-in-practice. The technique probably had its origin in the Buddhist monastic tradition. The mendicants likely picked up different healing methods and formulae during their wanderings. These were then collected and incorporated into the rule book for cenobites by putting the cures in the mouth of the founder, the Buddha. In the same way, additional medical information from healers that Buddhists came in contact with, such as Jīvaka, The Pediatrician, was stored in the same monastic rules. The codified medicine developed both inside and outside Buddhism and was the foundation on which early Ayurveda was built.

The case-by-case method of early Buddhism developed in early Ayurveda in an elaborate literature that divided therapeutics into diseases and their cures. Medical conditions were diagnosed in accordance with a nosology on the basis of a fixed number of pathogenic agents. This was followed by a prescribed treatment regimen with both physical treatment and pharmaceutical recipes. The whole of the early Ayurveda corpora was used for teaching and reference. As part of the priestly Brahmanic tradition, these medical textbooks were intended to be memorized by men trained to be physicians. In the beginning, it was an apprenticeship type of education, often passed on through families of physicians, but more recently, education has become institutionalized, where the medicine is frequently represented as a blending of biomedicine and Ayurveda. Today, few, if any, students memorize

6 Zysk (2021b).

Ayurvedic classics, but they are taught to revere them and are made to learn a few key verses by heart.

Although ancient Indian medical history remains obscure in its early period, there appear to be largely indigenous forces at work. While some impetus probably also came from the outside, the overall organizational principles, theoretical framework and methods of treatment are derived from indigenous doctors and healers. The crucial transition phase of ancient Indian medical history from Vedic to Ayurvedic medicine, like many others at about the same time, is the result of the dynamic process of both adaptation and innovation.

4.12. Appendix

Buddha's list of diseases treated

The list is the first attempt at systematizing medical diagnosis and treatment, and is presented here in the order found in the text:

1) large sores;

2) non-human disease;

3) eye disease;

4) head irritated by heat (head disease);

5) affliction of wind;

6) wind in the limbs;

7) wind in the joints;

8) split-open feet;

9) swellings;

10) snakebite;

11) the effects of harmful drink (two types of poisons);

12) defective digestion;

13) morbid pallor (jaundice);

14) corruption of the skin;

15) body filled with the pathogenic agents;

16) wind in the abdomen;

17) burning in the body;

18) rectal fistula.

4.13. Sources and recommended reading

The best bibliographical source is Dominik Wujastyk's 2011 contribution "Indian Medicine" to the *Oxford Bibliographies Online (https://www.oxfordbibliographies.com/view/document/obo-9780195399318/obo-9780195399318-0035.xml?rskey=eAcL50&result=1&q=Dominik+Wujastyk#firstMatch).*

Additional reading

Salguero, C.P. (2017). *Buddhism and Medicine. An Anthology of Premodern Sources.* Columbia University Press, New York[7].

Zysk, K.G. (2021a). Doṣas by the numbers. Buddhist contributions to the origins of the Tridoṣa-theory in early Indian Medical Literature with comparisons to early Greek theories of the humours. *History of Science in South Asia*, 9, 1–21. doi: 10.18732/HSSA68.

Zysk, K.G. (2021b). *Asceticism and Healing in Ancient India. Medicine in the Buddhist Monastery*, 2nd edition. Motilal Banarsidass Publishing House [Indian Medical and Ayurvedic Heritage], New Delhi.

Zysk, K.G. (2022). Note to 'Doṣas by the Numbers'. "Suśruta Project" Blog, University of Alberta [Online]. Available at: https://sushrutaproject.org/2022/01/22/note-to-do%e1%b9%a3as-by-the-numbers/.

7 This work offers a glimpse into ideas about medicine and healing in Buddhist records, primarily from East Asia.

5

Greek Medicine: A Brief Historical Outline

Lorenzo PERILLI
Università degli Studi di Roma Tor Vergata, Rome, Italy

5.1. Historical outline

5.1.1. *Hippocrates and the beginning of Western medicine*

If there is a figure who has left an indelible mark on the history of Western science and culture, this is undoubtedly Hippocrates of Cos. Medicine, both ancient and modern, is molded around this name, whose fame is as great and admiration inexhaustible, as the historical data at our disposal are limited. What we call "Hippocratic medicine" quickly became a paradigm; it holds the place of honor among the bodies of knowledge that allow us to understand the Greek "scientific" environment of the 5th and 4th centuries BCE, as well as the broader social and cultural context of the time: the efforts made by the Hippocratics toward understanding the human body, its structure, functioning, interconnections and its relationship with the external environment had an impact that would prove enduring and profoundly influence both ancient culture and the many centuries that followed. Medicine is also the first discipline to have explicitly been the object of critical self-appraisal.

With the contribution of Jutta Kollesch and Diethard Nickel, *A Survey on the Medical Literature of Antiquity*, published by the Berlin-Brandenburg Academy of Sciences.

The crucial phase of this evolution is marked by a textual tradition that emerged in the 3rd century BCE Alexandria, initially consisting of 30 works that then doubled over time until finally becoming a corpus – the *Corpus Hippocraticum*.

Although these sources represent only a tiny sample of what must have originally been produced, medicine is the only branch of ancient Greek science for which a sizable number of complete works, whose origin can be traced back to the age of Pericles' Athens, have reached us. The general label of *Corpus Hippocraticum* is misleading: the illusion of unity it creates masks the fact that this collection includes texts that differ in terms of authorship, date, content and approach. Therefore, we should be aware that the label "Hippocratic" serves as a shorthand term for medicine of the classical age in general, including different physicians/authors and remarkably dissimilar perspectives.

Some of the treatises cataloged under the name of the father of medicine are strictly technical, concerning, for example, diseases or gynecological problems (such as the case histories collected in *Epidemics* or the treatises entitled *On diseases II, On affections, On internal affections, On the diseases of women* and *On the nature of women*). Often drafted in the form of collections of clinical reports, they provide a diagnosis but no general theoretical framework. Other texts instead present rationalism as the bedrock of medicine: rejecting traditional beliefs, they sometimes take a clear stance on philosophy and the method of research. Most notable among these latter works are some texts that display a striking methodological awareness: some scholars have argued that no one can claim to know ancient philosophy unless they have read *Ancient medicine*, a work attributed to Hippocrates. *On the sacred disease* should be added to this, namely a veritable manifesto of rationalism, which assigns a purely natural and rationally understandable character to epilepsy, a disorder whose origin was considered divine on account of its uncontrollable and dramatic manifestations.

Some works lay the foundations of medical semiotics by stressing the crucial importance of foreseeing the evolution of a disorder on the basis of signs and their development into symptoms (*Prognostic*); other texts, such as the sophistic *The (medical) art*, uphold the value and trustworthiness of medicine. Other striking pieces of writing include books I and III of the *Epidemics*, a collection of reports and observations recorded by a physician during his visit to patients in various places (this is also the meaning of the title); *On regimen in acute diseases*; and *On airs, waters, places*, a remarkable treatise that foreshadows the discipline of anthropology in its analysis of peoples and climates, and the features (or disorders) correlated with them. Essential deontological and ethical precepts are provided by later works, from the Hellenistic age or subsequent periods, including the *Oath*: to this day, many aspiring physicians still receive a copy of the *Oath* from their university upon graduation.

It is common knowledge that, by the time the so-called "Hippocratic" writings received their final form, a vast harvest of material on medical topics was circulating. This was considered common property and was often subject to rehash and expansion by physicians working in isolation or with others. Among the many manifestations in which medicine found expression, the most important alternative to dominant rationalism is healing practices in sacred places such as sanctuaries, especially those devoted to the patron god of the craft Asclepius (Aesculapius in Rome). Other figures, too, were connected to the practice of healing: the *pharmakopolai*, or drug-sellers; the *rhizotomoi*, or "root-cutters", experts in the foraging and use of medicinal herbs and plants; the *maiai*, women doctors working especially as midwives, but also filling other roles; priests and *iatromanteis*, namely, physicians/seers; and experts in practical issues called *cheirotechnai*, perhaps to be interpreted as surgeons in a passage of the tragic poet Sophocles. To this we should add a phenomenon common in all historical ages, namely, the exploitation of patients' credulity and yearning for recovery – something that the Hippocratic author of *On the sacred disease* labels quackery and charlatanry.

5.1.2. *The paradigm of rationality*

In the absence of anything comparable to today's chemistry and of a remedy against the risk of infection inevitably associated with surgery, although herbal and other remedies of some efficacy existed, ancient medicine possessed but a limited capacity for treatment. The most distinctive contribution made by what we refer to as Hippocratic science lies in its capacity to reflect on its own nature and methods by emphasizing the peculiarity of medical procedures and the need to clearly distinguish them from the "hypotheses" formulated by philosophical figures such as Empedocles. One of the most significant discoveries or intuitions of ancient medicine concerned the role of the brain in processing feelings and as a "messenger for intelligence". First outlined by Alcmaeon of Croton and then taken up by the philosopher Anaxagoras a few decades later, this theory – which was also admired by Plato – was finally abandoned in light of the criticism it received from Aristotle, whose authority long stifled any further development in this direction. The most explicit formulation of the theory, which shows full awareness and bears first-hand witness to the phenomena described, is to be found in the treatise *On the sacred disease*:

> The fact is that the cause of this affection, as of the more serious diseases generally, is the brain. The manner and the cause I will now

> set forth clearly. The brain of man, like that of all animals, is double, being parted down its centre by a thin membrane [...] Veins lead up to it from all the body, many of which are thin, while two are stout, one coming from the liver, the other from the spleen. The vein from the liver has the following character. One part of it stretches downwards on the right side, close by the kidney and the loin, to the inner part of the thigh, reaching down to the foot; it is called the hollow vein [...] By these veins we take in the greater part of our breath…
>
> I hold that the brain is the most powerful organ of the human body, for when it is healthy it is an interpreter to us of the phenomena caused by the air, as it is the air that gives it intelligence (*phronesis*). Eyes, ears, tongue, hands and feet act in accordance with the discernment of the brain; in fact the whole body participates in intelligence in proportion to its participation in air. As regards understanding (*synesis*), the brain is the messenger. For when a man draws breath into himself, the air first reaches the brain, and so is dispersed through the rest of the body, though it leaves in the brain its most active part, and all that it has of intelligence and sense. If it reached the body first and the brain afterwards, it would leave discernment in the flesh and the veins, and reach the brain hot, and not pure but mixed with the humour from flesh and bloody, so as to have lost its perfect nature. Wherefore I assert that the brain is the interpreter of understanding (Hippocrates, *On the sacred disease*, ch. 3–4, 16; transl. W.H.S. Jones, ch. 6–7, 19, with changes)

The central role played by inductive-deductive reasoning clearly emerges here and counts among the most enduring contributions made by Greek medicine of the classical age. First-hand observation of available data ensures the discovery of the sought after proof; however, far from stopping at this, it triggers a far-reaching process that assigns the brain a more general role of instrument for rational understanding.

Often, it is not so much a matter of diagnosis, or indeed treatment: some of the patients – the author of *Prognostic* explains – will die soon after falling ill, others shortly after having seen a doctor, who will immediately be able to tell which patients have any hope of surviving and are therefore worth treating. The crucial aim, then, is to practice prognosis. The physician addresses individual facts, *ekasta*, which become signs and hence symptoms (*semeia, tekmeria*). After setting these symptoms within a framework of knowledge through an inferential process (*tekmairesthai*), the doctor can start from them; however, he must also return to them in the moment in which his conclusions are to be transformed into a form of treatment targeting precisely those data – namely, the patient and their condition.

Unlike other disciplines, because of its peculiar field of enquiry and application, medicine has always been an object of discussion and not just on the part of physicians. Plato is a privileged witness to this, since he often uses medicine as an example. He takes up the theme of "studying the body by studying the nature of all things", which is explored in his *Phaedrus* (270c) and in the short dialog *Charmides*:

> You have probably heard this about good doctors, that if you go to them with a pain in the eyes, they are likely to say that they cannot undertake to cure the eyes by themselves, but that it will be necessary to treat the head at the same time if things are also to go well with the eyes. And again it would be very foolish to suppose that one could ever treat the head by itself without treating the whole body. In keeping with this principle, they plan a regime for the whole body… (Plato, *Charmides*, 156b-c, transl. R. Kent Sprague)

In the oldest and methodologically most significant works collected in the *Corpus Hippocraticum,* much attention has been given to the study of diseases and of the physiological and pathological mechanisms in humans, along with the awareness of the necessary epistemological resources.

5.1.3. *Beyond Hippocrates: from Southern Italy to Alexandria*

Aside from the works of the *Corpus Hippocraticum*, which have been transmitted anonymously (with the only problematic exception of *On the nature of man*, which Aristotle seems to attribute to Polybus, Hippocrates' son in law), other ideas, tendencies and approaches to medicine should be mentioned.

First and foremost was the so-called "school" of medicine from the southern regions of the Italian peninsula, which found its leading and most ancient representatives in Democedes of Croton, Alcmaeon of Croton, Philistio of Locris, Acron of Acragas and Hippon of Rhegium in the late 6th and early 5th centuries. Alcmaeon is reputed to have cultivated considerable medical and biological interests, and to have attained some peculiar insights. Philistio, Acron and Hippon shared the belief that medicine is based on the search for the *arche*: they sought to reduce multiple pathological conditions to a few starting conditions, an approach reminiscent of that of the Ionian school and those theories based on *hypothesis* which the author of *Ancient medicine* opposed. They based their ideas on the search for not so much the basic elements, but their qualities or *dynameis* (hot, cold, moist and dry), whose balance ensured good health according to Alcmaeon (who nonetheless posited an indefinite number of them, rather than just "one or two", as *Ancient medicine* polemically notes).

However, apart from the leading centers and figures, many physicians and healers were operating autonomously in centers of various sizes, and frequent – if meagre – evidence of their presence survives in the form of inscriptions, given that these doctors have left no writings or any other traces.

In the Hellenistic period, medicine started taking a new form. For the first time, medical theories and conceptions began to be associated with the names of their proponents, whereas anonymity had previously been the rule. As with other disciplines, Aristotle's school sought to develop a detailed framework for medicine: Meno wrote a treatise on the history of medicine, which was later given the title of *Menoneia*. This work, now lost, may have been the direct source for a 2nd-century CE medical doxographical treatise preserved by a British Library papyrus (Inv. 137). Known as Anonymus Londiniensis, the text consists of almost 2,000 lines, for a total of 39 columns divided into three parts. The first, which shows Stoic influences, is devoted to the definition of various concepts, not limited to the field of medicine (e.g. the notion of illness and that of science). The second part is devoted to the etiology of diseases, which is presented on the basis of a medical doxography that mentions a number of authors, including many who were unknown prior to the publication of the papyrus in 1893. The third part of the text, which is very peculiar and relevant for us, features an exposition of Plato's medical conceptions.

The 4th century BCE was a period of transition, in which the name of Diocles of Carystus stands out. Active in Athens in Plato's day, this *Hippocrates iunior* pursued the study of anatomy, physiology, embryology, surgery and dietetics. While he is regarded as one of the greatest physicians of Antiquity, only some fragments of his works survive. Relations between Diocles and both Aristotle and Theophrastus have been posited on the basis of certain theoretical affinities; however, although he was even believed to be a pupil of Aristotle, his name does not occur in the latter's corpus, nor does Aristotle's name ever occur in Diocles' fragments.

Fragments are all that survive even for almost all subsequent physicians, down to Galen, who marks the high point in Greek medicine. However, despite this dearth of material, we know that a crucial contribution was once again provided by the city of Alexandria, particularly by Herophilos of Chalcedon (Bithynia) and Erasistratus of Ceos (Cyclades), in the late 4th/early 3rd century BCE. These two figures are credited with a number of discoveries: Herophilos – a pupil of the influential Praxagoras of Kos (latter half of the 4th century BCE), an anatomist and supporter of the Hippocratic theory of humoral pathology – is regarded as the discoverer of the fourth ventricle and of the choroid plexuses; he is also known to have carried out detailed studies on the liver, duodenum, lungs, pulmonary artery, prostate, eye, heartbeat and nervous system – establishing the distinction between sensory and motor nerves. Herophilos developed a theory of pulsation as a diagnostic instrument. He was also interested in dreams and their interpretation, a topic that received much attention in Antiquity – and not just among the Greeks either. Herophilos has made innovative contributions to anatomy and physiology, whereas in the field of therapeutics and pathology his role was more that of a transmitter than an innovator.

Erasistratus' work, like that of Herophilos, reflects the influence of Aristotelianism. Erasistratus was the author of texts on anatomy, physiology, dietetics and pharmacology still available to Galen in their original versions, but which were later lost, except for some fragments. Like Herophilos, Erasistratus also practiced dissection and is credited with explaining the function of cardiac valves and the observation that the heart operates as a pump and has nothing to do with feelings – of which it was instead held to be the center by many philosophers. However, the mechanism of the circulation of blood, which he compared with the flow of water in an irrigation system, was not fully understood yet. Erasistratus was the first to distinguish between veins and arteries, but he regarded the latter as being filled with air and responsible for the circulation of *pneuma*, the vital spirit. This was one of the central elements in his conception of disease, which, in his view, was derived from the spilling of blood from veins into arteries, thus preventing the correct functioning of the *pneuma*. Erasistratus furthered the study of the nervous system and various kinds of nerves; he investigated the brain and its parts, also in their connection to nerves. His conception of the human body and its physiology betray the specific influence of Aristotelian philosophy, especially that of Strato of Lampsacus, who was probably the originator of the corpuscular conception of things as consisting of small *invisible* particles and voids, even in relation to the human body, although *horror vacui* here prevented the positing of empty spaces beyond a certain size.

Herophilos and Erasistratus present us with what may be regarded as genuine medical systems. Both operated in Alexandria, a city in which the enlightened support of the Ptolemies ensured the application of Aristotle's scientific program. Here, also in the field of mathematics and especially astronomy, a real tradition soon took root. The contribution of Hellenistic medicine at this stage chiefly consisted of the introduction of detailed anatomical studies through the dissection of bodies not only of animals, but also of humans (which previously had seldom been used). Vivisection was also performed on prisoners on the death row, whom the authorities apparently made available for medical research. This led to a far more detailed knowledge of the body, its organs and their functions, despite the endurance of many uncertainties and unproven hypotheses; on the methodological level, the advancement was certainly remarkable. Nevertheless, these developments remained isolated: soon afterwards, with the birth of the various "schools", or rather approaches, of the Hellenistic and Imperial age – the Empirical, Methodist and Dogmatic schools – postmortem anatomical research was soon abandoned, not only in practice, but also through arguments intended to prove its uselessness. Christian thought (e.g. Tertullian) further contributed to the denigration of the Herophilean dissection of the human body; the fact that Herophilos had devised an instrument for the destruction, under extreme circumstances, of the fetus in its mother's uterus – the *embryoclast* – was a further reason for the controversy.

The rejection of anatomy and, in the case of the Empiricists, physiology was based on preconceptions: it amounted to a rejection of all causal systems and the very idea of the unitary nature of the human body and hence of the possibility of acquiring direct knowledge of it. Alongside these were some more specific motivations: on the one hand, the remarkable discoveries made by Herophilos and Erasistratus had not led to any significant progress at the therapeutic level; on the other hand, it was argued that the conclusions reached by studying a corpse could not be extended to a living body, as the two conditions were too different.

5.1.4. *Hellenistic medical schools and their controversies*

Once again, medicine experienced a setback. This was not the case with medical debate, however, which raged on with the three aforementioned Hellenistic "schools". As the original works by their exponents are lost, Galen is the primary source on the subject. Notably, what is meant here by "school" is not an actual institution but a shared ideological approach to science. The Hellenistic and Imperial schools fueled a lively methodological and philosophical debate rather than making significant progress in either therapy or diagnostics.

Under the labels "Empiricist", "Methodist" and "Rationalist" (or Dogmatic), with the later addition of the pneumatist "school" and of individual physicians who described themselves as "Democritean" or "Erasistratean", the representatives of the different currents of medicine came together, or were brought together by later interpreters, starting from Herophilos' disciples, and especially Philinus of Cos. The latter is regarded as the founder of empiricism around the year 260 BCE, although according to some testimonies, this school should be traced as far back as Empedocles' pupil Acron of Acragas in the 5th century BCE. Shortly afterward, other representatives were Serapion of Alexandria and, later, Menodotus of Nicomedia (2nd century CE), marking the culmination of the school just before

Galen. Heraclides of Tarentum (1st century BCE) has also been considered as an adherent of this medical approach, although he adopted a more personal approach and even criticized certain aspects of empiricism.

The label "Rationalist" is applied to nearly all of the chief exponents of medicine from the Classical Age onward, including Hippocrates himself. Instead, Empiricism and Methodism represent the ultimate development of two crucial aspects of medicine: the centrality of experience and that of rationality. A radical conflict and violent polemic emerged between the two currents: Galen, our main source of information, emphasizes the impact of this contrast and, in some cases, even the danger it constituted on a social as well as scientific level. Empiricism, the most long-lived of the Hellenistic medical schools, abandoned the idea of searching for the causes of diseases as a pointless endeavor: all that mattered, for this school, was the actual treatment of patients, essentially on the basis of the physician's previous experiences.

Methodism (so called because of its proponents' aspiration to follow a single method) is chiefly associated with Themison of Laodicea (latter half of the 1st century BCE) and Thessalus of Tralles (1st century CE), although the origins of the school are traced back to Asclepiades of Bithynia, in the 1st century BCE. Both the Empiricists, especially Menodotus, and Galen engaged in a vehement contention with the Methodists. Galen revealingly describes figures such as Asclepiades and his later followers as "so unintelligent and slow-witted that even by the time they have reached old age they are still incapable of understanding the steps of an argument": people of the sort that once "used to be set to menial tasks. What will be the end of it, God knows!" (Galen, *On the natural faculties* II, 53,5, Kühn).

As the exponent of an atomistic doctrine founded on an idea of corpuscles whose existence and function were posited not on the basis of observation, but of sheer rational speculation, Asclepiades posited that inner disorders are due to an alteration in the flow of atoms through channels or pores in the human body, which in turn was believed to be formed by chains of interlocked atoms. Themison and Thessalus took these channels to be phenomenal entities whose dilation or constriction could alter the balance of the body, leading to illness. The exponents of rationalism instead affirmed the priority of the rational element, giving indications on the basis of experience and finally yielding knowledge. For the empiricists, by contrast, priority was assigned to experience and autoptic examination according to the various modes of observation.

Galen's aversion to the Methodists also had a more specific target: the claim that complete mastery of the medical art required not years or indeed a whole life of study, as Galen himself argued, but only six months – a claim that was all the more dangerous, given the wide popularity enjoyed by Methodism.

5.1.5. *Awaiting Galen: Rufus and Soranus*

The path leading to Galen is dotted with the names of noteworthy figures, who contributed to medical practice, diagnosis and therapeutics. One of the most influential figures, a physician greatly praised by Galen himself, was Rufus of Ephesus (c. 100 CE). Operating in Egypt and Rome, he resumed the Alexandrian study of anatomy: in the case of man, he apparently limited himself to external anatomy but employed monkeys for studying the internal workings of the human body, as these had long been recognized to be the animals most similar to humans and therefore most suitable for anatomical research. Rufus' pragmatic approach was undoubtedly one of the factors that contributed to his fame. A degree of emotional involvement in patients' case histories is detectable in his surviving writings, along with an interest in socially marginal groups such as slaves. Rufus' doctrines circulated in the Arab world, which has yielded translations of important works, the original versions of which are lost. Most notable among these is *Case histories*, a sort of clinical diary of cases treated. Rufus regarded himself as Hippocratic and was interested in humoral pathology. He is credited with a number of discoveries related to the anatomy of the eye, the uterine tubes and the pancreas. In his writings, Rufus also focused on various aspects related to actual medical practice (for instance, in his study of the kidneys and bladder), abandoning the discussion of philosophical issues, which remained a typical feature of many medical treatises at the time. One work that stands out is *On the names of the parts of the human body*.

At about the same age, another outstanding physician, Soranus of Ephesus (1st century–2nd century CE), started practicing in Rome after completing his studies in Alexandria. Regarded as an exponent of the Methodist school, even though he did not strictly confine himself to its principles, Soranus is known as the founder of gynecology and scientific obstetrics – his work provides the first surviving description of the subject. Soranus' treatise *Gynaecology* remained a point of reference until the Renaissance, along with other technical writings by him on both related (e.g. *On the uterus and female genitalia*) and unrelated subjects (e.g. *On signs of fractures* and *On bandages*), which survive in fragments. In addition, Soranus wrote some philosophical works as well as commentaries on Hippocrates' teachings, now lost. His main work, a treatise on acute and chronic diseases, has been preserved through a Latin version produced by Caelius Aurelianus in the 5th century CE.

5.1.6. *Galen: intellectual, philosopher, prince of physicians*

Galen of Pergamon (129–c. 216 CE) exercised an enormous influence as a physician and philosopher, only comparable perhaps to that exercised by Ptolemy in the field of astronomy. Galen's conception of medical science and the importance he assigned to Hippocrates influenced the ancient, Arab and Byzantine worlds for centuries; through Galenism, they remained central until the development of modern medicine.

Following studies in Pergamum and Alexandria, Galen arrived in Rome during the rule of Marcus Aurelius Antoninus, the emperor philosopher, in the 2nd century CE. Here, he declared Hippocrates to be his hero, considering himself his heir; although not even a single treatise could be attributed to the father of medicine with certainty, he strove to exhalt his image and build a construct of what medicine should be. In this Galen was remarkably successful. His claims were so forceful and convincing, his arguments were so assertive, and his criticism of adversaries was so sharp and penetrating that his ideas influenced the evolution of Western medicine until the 19th century. Galen was the new Hippocrates, and as such, he came to represent the ideal physician.

Galen's medicine must be seen as the crowning achievement of a seven-century-long history and as the basis for teaching and practice throughout Europe. He was effective in depicting himself as the best of all the doctors, and his knowledge of his predecessors' works, exceptional practical ability and wide-ranging clinical experience went a long way in supporting his claim. His own education was unprecedented, and with him, medicine underwent a revolution of concepts that resulted in new paradigms that drastically altered the practical features of healing activity throughout Antiquity and later centuries.

The corpus of Galen's writings is more extensive than that of any ancient author. Unlike others, many of his works are still available, either in their original Greek version or in Latin, Arabic and even Syriac and Jewish translations. Approximately 140 of these texts survive, whereas the complete list of transmitted works – including many regarded as spurious – comes to a total of approximately 400 titles. These are equally divided into treatises on medical subjects (anatomy, pathology, therapeutics, diagnostics and prognostics) and on philosophy, logic and philology, including works on language. A separate category is represented by Galen's valuable and extensive commentaries on Hippocrates. The corpus further includes some unique works for Antiquity, such as the detailed autobiography *On My Own Books*, and what has been described as a philosophical testament, *On My Own*

Opinions. Besides, it was as a philosopher that Galen wished to be remembered, and his contribution no doubt deserves a place in any history of ancient thought.

According to Galen himself, Marcus Aurelius described him as "first among physicians", "unique among philosophers" and a "free man" in the philosophical sense (*On Prognosis – De praecognitione*, 11,8). What is rather surprising and has yet to be adequately explained is the fact that the literature contemporary or immediately subsequent to Galen makes only fleeting mentions of him. It seems as though Galen had no direct disciples or successors, although he favored a didactic approach, as several of his works suggest. "Galenism" only emerged in the 4th century with Oribasius, the first of the great medical compilers of Antiquity, at a time in which the drafting of original works had already come to be regarded as a superfluous pursuit, given the availability of an extensive corpus of medical knowledge, ranging from Hippocrates to Galen, which only needed to be rearranged.

The son of an architect, Galen received thorough philosophical, mathematical and medical education. He spent five years in Alexandria, in addition to Smyrna and Corinth, before moving to Rome. Here, he soon became the physician of the Roman imperial court, chiefly under Marcus Aurelius (but later also under Commodus and Septimius Severus). Galen sought to apply the Euclidean criteria of demonstrative rigor and consistency to the field of medicine, which are based on axioms and definitions, so as to establish a genuine scientific system that would lend a unitary character and logical-philosophical coherence to the body of Greek medical knowledge. With this aim in view, Galen also drafted specific works, such as the logical treatise *On demonstration*, which were conceived as preliminary reading to any serious education in the field.

Unlike many other physicians of his day, Galen achieved considerable therapeutic success – so much so that he came to be regarded as a sort of miracle worker. In his years of study in Alexandria, he acquired extensive anatomical knowledge, which he further increased by working as a gladiators' physician. Galen accepted and contributed to perfecting the ancient Hippocratic doctrine of humors, which was based on the idea of a balance between four fundamental liquids in the human body (blood, phlegm, yellow bile and black bile), corresponding to four main temperaments (sanguine, phlegmatic, choleric and melancholic). He conceived disease as an alteration of this balance or as an imbalance between the four fundamental qualities. Alterations in the combination of the four humors, and hence of a person's temperament, were seen as the cause of disorders. Unlike the representatives of Hellenistic-Imperial schools, Galen believed that medical knowledge also extended to the four fundamental elements of the human body (which others regarded as the object of natural philosophy): air, water, earth and fire, with the corresponding qualities being cold, moist, dry and warm. Medicine, therefore, came to replace the philosophy of nature, not least because it could lay claim to a far more reliable method of demonstration.

Galen attempted a medical-philosophical syncretism that brought together the venerated figures of Hippocrates and Plato, the best philosopher (although this verdict did not prevent Galen from incorporating a hefty dose of Aristotelianism into his work): the two names exemplify Galen's attempt to re-establish the unity of science by overcoming those divergences on which Empiricists and Methodists had been arguing for a long time with no benefit for medicine. Galen combined methodological elaboration and empirical verification, the latter constituting both the starting point and testing ground for the former. Accordingly, Galen often resorted to experiments, many of which were conducted before the public, not least to measure himself against his rivals and assert his superiority in a competitive social context where the absence of any official medical training and the wide range of often unreliable practitioners often led to disputes.

Medicine, for Galen, was meant to demonstrate what philosophy affirmed: he sought to provide physiological and anatomical confirmation of the Stoic philosophers' idea of *pronoia*, namely, a providential universal order. In his view, the perfect mechanism governing a healthy body was proof of this finalism, which in turn explained why each organ has its own natural faculty, determining its specific function. According to such a view, nature has provided the body with organs and functions that stand in a reciprocal relationship with one another to ensure the orderly functioning of the whole. The latter is altered by those pathological states that the physician seeks to investigate and bring back to the original condition. Medicine therefore came to be an instrument for the rational investigation of the providential nature of the world. Unsurprisingly, Galen's reflections on the matter reveal substantial affinities with the arguments of early Christianity.

Galen shared the Alexandrian knowledge of the nervous, venous and arterial systems. He regarded the nerves as channels filled with psychic pneuma – a fluid derived from the air one breathes – and divided them into sensory and motor: the former carrying perceptual stimuli to the brain and the latter regulating voluntary movement by conveying orders from the brain. The nerves were connected to the brain and spinal marrow. Finally, abandoning the Aristotelian idea of the centrality of the heart, which hindered the study of the brain because of the authoritativeness of the master, Galen asserted that the latter, and not the heart, was the seat of the rational part of the Platonic soul, governing higher psychic functions. The other central organ in his view was the liver: in contrast to what Alexandrian medicine had suggested, Galen described it as the origin of veins and blood. Galen explicitly drew upon the tripartite conception of the soul, chiefly presented by Plato in the *Timaeus*: the brain was seen as the seat of the rational soul, the heart as that of the passionate soul and the liver as the seat of the nutritive and reproductive soul, also governing bodily desires.

As already noted, Galen composed works ranging across several fields of knowledge. An uncommon figure, particularly in the field of science, Galen also reveals himself in the genuineness that emerges in *On the absence of grief* (*De indolentia*), where he describes the loss of most of his books in the fire that destroyed Palatine Hill in Rome in 192 CE, highlighting his cultural and intellectual sensitivity. Other works, by contrast, reveal his more ambitious, self-celebratory and polemical side, sometimes verging on unfairness. Galen's main and most influential works include *Methodus medendi* ("On the method of healing"), in 14 books transmitted by the late medieval tradition as *Megategni* (from the Greek *mega techne*, great art), and the shorter *Ars medica*, later known as *Tegni*. Other remarkably rich works are his pharmacological treatises, such as *On the mixing and the properties of simple drugs*, as well as his works on anatomy and physiology, such as *Anatomical procedures, On the usefulness of the parts of the body* and *On the natural faculties*, which count among his most important texts. To these we may add Galen's treatise on arteries and their content, *Whether arteries contain blood; On the dissection of nerves* and *On the dissection of veins and arteries*; more markedly philosophical works, such as the *Institutio logica* and *On demonstration*, a (now fragmentary) text on logic; *On the doctrines of Hippocrates and Plato*, *On the elements according to Hippocrates*, and *That the best physician is also a philosopher*; last but not least come the aforementioned autobiographical and autobibliographical treatises, for which no parallel can be found in the ancient world before St. Augustine.

With Galen and Ptolemy, names that soon acquired symbolic value, the 2nd century marked the end of the most creative phase in ancient science.

5.1.7. *Galen's aftermath: the journey from Rome to China*

Soon after his death, Galen was largely neglected, and despite his prodigious literary output, his contributions, with very few exceptions, were rarely mentioned. He remained known and respected, as confirmed by the scant existing references, but his contemporaries mainly behaved as if he had never existed. It is extraordinary that in a discipline in which oral transmission of knowledge was essential, he had – at least according to the available information – no pupils or followers. All in all, for a doctor who said of himself "great was the name of Galen!" (*On prognosis*, 5,4), we would expect a much wider resonance than what we find in the historical record and at least some discussion of his theories and results. Among the factors responsible for this period of oblivion, two appear likely. On the one hand, Galen's authority may not have been acknowledged beyond small learned circles. On the other hand, the status of a physician in Rome was ill defined, as the practice of medicine was not considered appropriate for a Roman citizen and was usually the domain of immigrants and slaves. It is probably not by chance that Alexander of Aphrodisias, a professor of philosophy at Athens, mentions him as a philosopher rather than a doctor.

In Late Antiquity (from the 4th to the 6th century), medicine became the domain of compilators, figures such as those of Oribasius of Pergamum, Aetius of Amida and Paul of Aegina. By selecting and collecting material from the treasure trove of the past, they intended to continue a tradition of practice rather than theory, sorting out recipes and remedies. Despite the variety of treatises that had been circulating for ten centuries, Galen and his system resurfaced to prevail over all others as the end-all of medical wisdom, a process termed *Galenism,* which was to last for centuries. Galen became the main source of later compilations, but the development of medicine was halted. The Hippocratic canon, acknowledged in Hellenistic times, became reality: "Medicine has been already discovered in its entirety", the author of *On the (medical) art* had once said in the 5th century BCE. No further autonomous research or innovation seemed to be needed. Galen's many books, his mastery of physiology, pathology, therapeutics and all other areas of medicine, fostered the idea that they contained all that was required for effective medical practice and for the education of aspiring physicians.

In the first decades of the 6th century, Hippocratic medicine was collected into a body of chosen treatises, whereas 16 works of Galen's were adopted as a canon to be read and commented upon following a sequence advocated by him. As knowledge of Greek was gradually declining, commentaries written by scholars were in Latin, which is now the language of medicine. Translations became essential to gain access to the great works of the past. Latin and Syriac versions began to appear as early as the 5th century CE (perhaps even earlier), when some Hippocratic treatises – such as *Airs, waters places, Aphorisms, On regimen* – and from Galen – *On sects in medicine, Art of medicine, On the pulse for beginners, Therapeutics to Glaukon*, which was the first work of the "Alexandrian Canon" – found their way into the social and cultural setting of the time.

As a culmination of this process of selection, reuse and dissemination, in the 9th century, a translation movement began in Baghdad under the guidance of Ḥunain ibn Ishāq, a physician and scholar responsible for numerous highly accurate translations of Galen's works into Syriac and Arabic. We also owe Ḥunain for a detailed index written in the form of a letter, the *Risāla*, listing the available Greek manuscripts and translations of Galen into both Syriac and Arabic. The index lists 128 works of Galen, each with the corresponding translation(s), translator's name and details of the quest for each original, as well as the technique of translation. Ḥunain's translations are among the most important factors responsible for the introduction of Greek medicine and science in the Arab world, and for its later rebound westwards by way of translations from Arabic into Latin.

Did the cross-cultural fertilization of the Arabic and Latin-speaking world ever reach China? The great medical work *Firdaus al-Ḥikmah* (The Garden of Wisdom), written in Arabic in the 9th century, contains remarkable quotations of Indian physicians intertwined with texts from Hippocrates, Galen and Dioscorides, and is an eloquent example of cross-cultural fertilization. Furthermore, a very unusual but meaningful account further confirms the existence of eastward transmissions of Greek medicine. The *Index of the Sciences* (*Fihrist al-'Ulūm*) of Abū' l-Faraj Muḥammad ibn Isḥāq al-Warrāq, known as Ibn Abī Ya'qūb al-Nadīm – a bibliographical catalog written in the 10th century – contains a story concerning the great Persian physician and alchemist Rhazes. After some remarks on how difficult and time consuming it was to write using standard Chinese script, the author explains that the process may be expedited with a technique known as collective writing, wherein a single character can be used for words that usually require three or more words. "If they wish", al-Nadīm added, "they (i.e. *Chinese scholars*) can write the contents of 100 leaves with this script on a single page". He explains further:

> Muhammad ibn Zakariyâ ar-Râzï said: a man from China came to seek me and dwelt with me for about a year. […] When he desired to return to his country, he said to me a month in advance: 'I am about to set forth and wish that you would dictate to me the sixteen books of Galen, so that I can write them down'. I said, 'Your time is short and the length of your stay will be sufficient for you to copy only a small part of it'. Then the young man said, 'I ask you to devote yourself to me for the length of my stay and to dictate to me as fast as you can. I will keep up with you in writing'. [...] We have a form of writing known as Collective, which is what you see. If we wish to write a great deal in a short time, we write it with this script (*Kitab al-Fihrist*, mit Anmerkungen herausgegeben von G. Flügel, 1–2, Leipzig 1871, p. 17 (Arabic text), transl. B. Dodge in *The Fihrist of al-Nadīm*, New York, 1970, p. 31).

Thus, a translation of Galen into Chinese could have circulated since the 10th century. Although no perceptible direct influence of Greek medical theories in

Chinese medicine has been identified yet, the anecdote is an example of those cross-cultural links that may have influenced the transformation and the spread of knowledge and Greek medicine to distant regions and cultures.

5.2. A survey on the medical literature of antiquity

The extract that follows is taken from *The Medical Literature of Antiquity*[1] by Jutta Kollesch and Diethard Nickel, published by the Berlin-Brandenburg Academy of Sciences.

Figure 5.1. *Hippocrates of Cos, Roman copy of an original from the Hellenistic period. Museum in Ostia. Illustration: DAI Rome; Inst. Neg. No.: 68.1499. Photo: Felbermeyer*

Ancient medical texts serve as sources for research into the history of science and culture, and as examples of Greek and Roman literature. Their prose style represents an important paradigm of specialist writing. The oldest fully extant works are those found in the collection of writings attributed to the physician Hippocrates of Cos (c. 460–370 BCE). The majority of the texts, around 60 in all, date back to the 5th and 4th centuries BCE. Written in the Ionian dialect, the texts demonstrate a

1 English translation by J. Crone. Published in: Brochure from the Corpus Medicorum Graecorum/Latinorum, Berlin 2004. Available at: https://cmg.bbaw.de/> Survey > The medical literature of Antiquity. © Berlin-Brandenburg Academy of Sciences. Published with permission.

wide variety of literary forms, deal with a wealth of subjects and exhibit a diversity of theoretical concepts.

The oldest items in the Hippocratic collection include records containing information on the symptoms and course of diseases ("Epidemics", books I and III), a treatise on the meaning of signs for medical practice ("Prognosticon") and two works dealing with bone surgery ("On Joints" and "On Fractures"), which testify to sound knowledge of the human skeletal structure. In the essay "On the Nature of Man", the classical form of the doctrine of the four humors is expounded, and the humors – blood, phlegm, yellow and black bile – are coordinated to the primary qualities wet/warm, wet/cold, dry/warm, dry/cold, as well as to the four seasons, i.e., spring, winter, summer and fall, respectively. As a physiological conception, the theory that health consists in the equilibrium of the humors, and disease in the disturbance of the mixture retained its significance for medical thinking throughout antiquity and far beyond. The treatise "Airs, Waters and Places" deals with the influence of environmental factors on human health and illness.

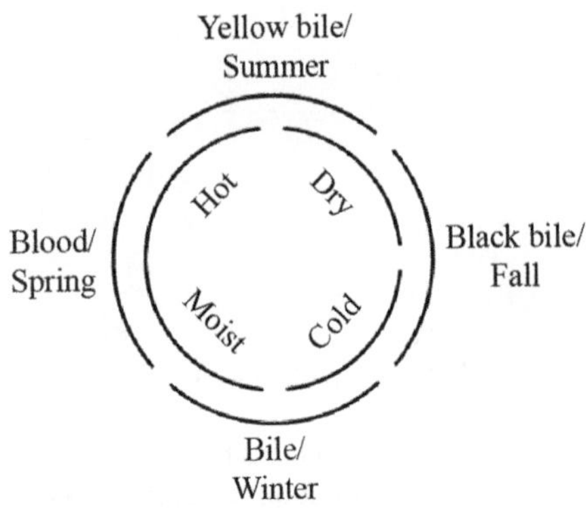

Figure 5.2. *The fourfold schema according to Hippocrates*

A characteristic feature of Hippocratic medicine is the view that healthy and pathological processes in the human body are the result of natural conditions, and that diseases should be treated by rational means. The Hippocratic physician considers the patient an equal partner whose dignity is to be protected in all respects. The successful treatment of a disease requires cooperation based on trust between the physician and patient. The ethical principles of Hippocratic medicine remain relevant even today.

Apart from a few texts belonging to the Hippocratic corpus that were not written until the Hellenistic period, the medical literature of this period has been lost almost entirely. That the literature of the Hellenistic period must have been extensive and substantial is clear from the frequent use made of it by later authors. Their reception of the Hellenistic legacy reveals that in the previous period, particularly in the specialized areas of anatomy, physiology, surgery and pharmacology, considerable progress had been made.

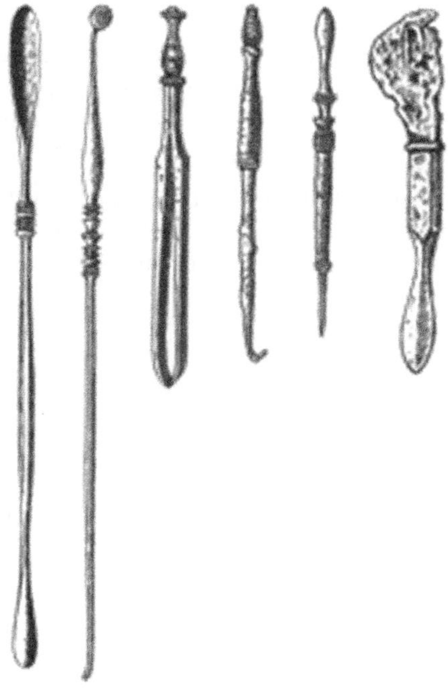

Figure 5.3. *Medical instruments from Pompeii (1st century CE). From left to right: Spoon probe, ear probe, forceps, retractor, couching needle and scalpel*

The establishment of individual medical disciplines is reflected in transmitted titles of lost works. Dating from the first half of the 3rd century BCE, books entitled "Anatomical Studies", "On Pulse" and "On the Eyes" are attributed to Herophilus of Chalcedon, whereas Erasistratus of Ceos is credited with "On Fever", "On Injuries" and "Hygienics". In the 1st century BCE, Crateuas wrote an illustrated herbal that has not survived too. However, an illustrated commentary on the Hippocratic text "On Joints", written by Apollonius of Citium, dating from the same period and continuing the Hellenistic tradition of Hippocratic exegesis, has survived intact.

In the Alexandria of the 3rd century BCE, physicians took advantage of the opportunity – without precedent in the ancient world – to conduct anatomical research on the basis of dissection of the human body. These procedures allowed errors to be corrected, which had arisen due to the fact that findings based on dissection performed on animals had been applied uncritically to humans. The more precise knowledge of the location and structure of bodily organs in humans that resulted from the procedures had a positive effect on medical practice, especially on operative surgery.

Largely based on Hellenistic sources and comprising eight books, "On Medicine" by Aulus Cornelius Celsus was probably composed during the reign of Emperor Tiberius (14–37). This work made the achievements of Greek scientific medicine available to Latin-speaking Romans for the first time. "On Medicine" is the only fully extant portion of an encyclopedia which also featured treatises on agriculture, rhetoric, philosophy, military and law. Of the remaining treatises, only fragments dealing with agriculture and rhetoric have survived.

Figure 5.4. *Depiction of an iris. From the work "On Pharmacology" by Dioscorides. Vindobonensis medicus Gr. 1 (6th century). Illustration: ÖNB Bildarchiv, Vienna. For a color version of this figure, see www.iste.co.uk/perilli/history1.zip*

The fact that the aforementioned subject areas were dealt with on an equal footing as part of an encyclopedic work constitutes an argument against their author's having been a physician. The medical field is geared toward practical medical requirements, and possible treatment methods are distributed among the fields of dietetics, pharmacotherapy and surgery.

A few decades later, the physician Dioscorides, a native of Anazarbus in Asia Minor, composed the five-volume "On Pharmacology", the first systematic presentation of the subject available to us. Although the author has clearly made use of the relevant literature from the Hellenistic period, much of the work is nevertheless based on firsthand experience. The explanations consist of descriptions of drugs from the three kingdoms of nature, as well as their medicinal properties and applications. In the later tradition, the arrangement of material in the text was modified, and high-quality artistic illustrations were added.

The oldest surviving anatomical writing is a textbook entitled "On the Names of the Parts of the Human Body", written by Rufus of Ephesus (c. 100). The main purpose of this work was to disseminate knowledge of anatomical nomenclature. However, the brief explanations also contain information on the location, shape and function of organs. Of particular interest is Rufus' essay "The Interrogation of Patients", as it is, to our knowledge, the only ancient text that deals specifically with medical anamnesis.

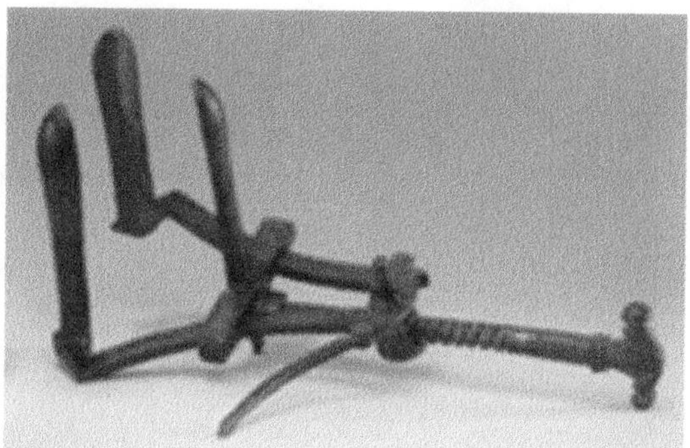

Figure 5.5. *Speculum, a device for spreading the vagina to facilitate examination of the uterus (1st/2nd century). Illustration: Römisch-Germanisches Zentralmuseum, Mainz. For a color version of this figure, see www.iste.co.uk/perilli/history1.zip*

One of the most important works of Rufus' somewhat younger contemporary, Soranus of Ephesus (c. 120), has been passed down in the Greek original. The work, entitled "Gynaecology", served up to the Middle Ages as the basis for gynaecological knowledge. The author also includes obstetrics, embryology, and infant care in the work. His textbook on acute and chronic diseases, however, has only survived in the

Latin version of Caelius Aurelianus, dating from the beginning of the 5th century. Finally, a short treatise "On Bandages", with accompanying illustrations, has survived in Greek.

In scope and content, the extant works of the physician Galen of Pergamum (129 to c. 215) constitute, together with the collection of Hippocratic writings, the most important contribution to ancient medical literature. In his work, independent medical research in antiquity reached its conclusion. The eclectic viewpoint he adopted allowed him to critically evaluate the achievements of the past and to integrate them into a comprehensive medical system. Consequently, in Galen's works, valuable ideas contained in the lost writings of his predecessors are preserved, perhaps no longer in their authentic form, owing to his high-handed treatment of the material available to him.

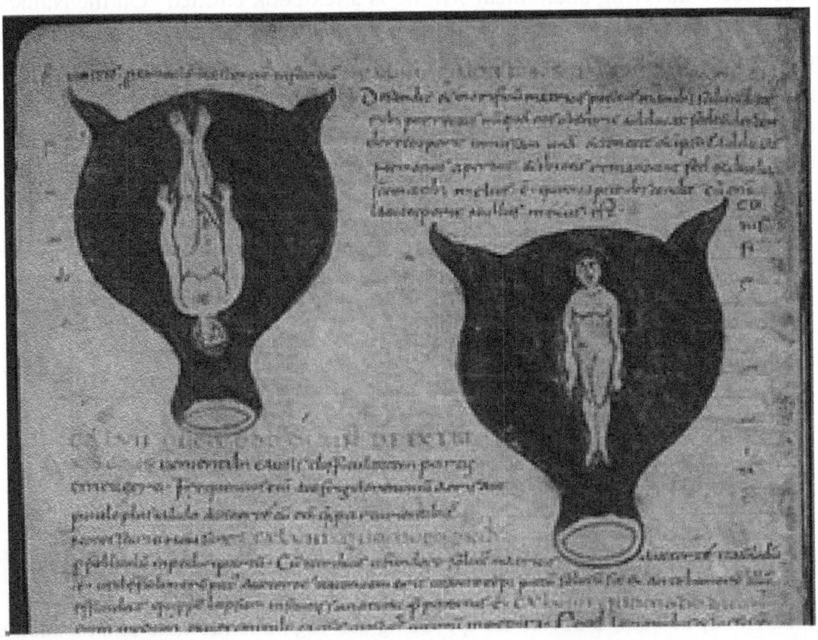

Figure 5.6. *Various foetal positions (head-first and feet-first). Illustrations to a late Latin translation of the "Gynaecology" of Soranus. Bruxellensis Lat. 3701–3715, fol. 26v (9th–10th centuries). Illustration: Copyright Royal Library of Belgium. For a color version of this figure, see www.iste.co.uk/perilli/history1.zip*

Galen occupied himself to varying degrees with nearly all medical specializations. He made independent contributions in many fields, but especially in anatomy, physiology, nosology and pharmacology. His physiology and nosology were based on the Hippocratic theory of the four humors, which he further developed in connection with Aristotle's theory of the four elements, coordinating the four humors of blood, phlegm, yellow and black bile to the four elements of air, water, fire and earth, which were held to be the basic elements of all existing things (the primary source for this is the treatise "On the Elements According to Hippocrates"). The connecting link consisted of the coordinating of the same combination of two primary qualities to one humor and one element. This expansion of the fourfold schema opened possibilities for the further development of the principles of Hippocratic medicine within the framework of Galen's humoral-biological theory of disease.

The doctrine of the four primary qualities and the four elements also served as the theoretical basis of Galen's pharmacology, which he outlined in a treatise entitled "On the Mixture and the Properties of Simple Drugs". Because he considered drugs to be composed of the same element-based primary qualities as the human body, Galen was able to integrate pharmacology into a self-contained medical system. The property of a drug was derived, in his opinion, from the primary qualities prevailing in it. He classified these qualities with regard to the intensity of their properties into four degrees, subdividing each degree into three levels.

In accordance with his principle that only a unified approach to medical research combining empirical observation with theoretical understanding can succeed, Galen, in his treatise "On the Usefulness of the Parts of the Body", elected a teleological approach to his subject matter, in that the goal of his presentation of anatomical structures was to prove their functional suitability. His knowledge of the anatomical structure of body parts was derived from dissections, which, as is evident from works such as "On Anatomical Procedures", he performed with a high degree of technical perfection on mammals. His depiction of the nervous system is based partly on vivisections, which aided him in demonstrating the presence of nervous system deficiencies through the interruption of nerve strands.

Hippocratic exegesis comprises a large portion of Galen's writings. In his exegetic works, he carries on the tradition begun in the Hellenistic period of improving contemporary understanding of Hippocratic texts, which were regarded as classic. By his own account, he commented only on those writings which he held to be authentic. These include, e.g., "Aphorisms", books I–III and VI of "Epidemics", the treatise "On the Nature of Man", and works on bone surgery.

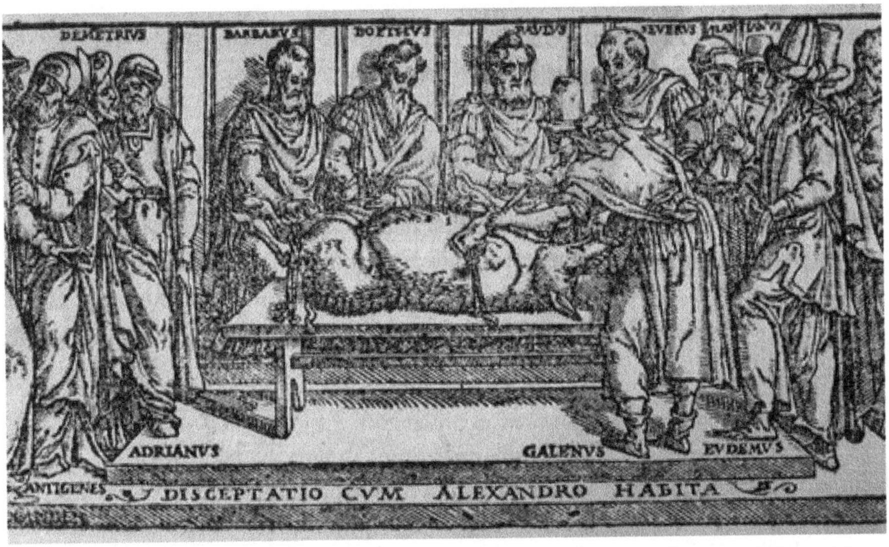

Figure 5.7. *Galen performing a vivisection. Illustration (detail) from the cover page of the 9th Galen–Juntina, Venice 1625. Illustration: CMG*

In late antiquity and the early Byzantine period, the medical authors of Greek-speaking areas were primarily concerned with summarizing knowledge gained in the past in a structured manner in order to preserve it for practical medical use. The resultant handbooks are also valuable from a scientific perspective due to their use of excerpts of previous texts, the originals of which are no longer extant.

The oldest of these works is the 70-book "Collection of Medical Texts" composed by Oribasius, who worked as an imperial physician at the court in Constantinople in the middle of the 4th century. In the 6th century, Aetius of Amida wrote a similar compilation in 16 books, which included magical practices among the therapeutic measures described. The list of medical handbooks written for practical purposes also includes a 12-book work in which Alexander of Tralles, working at around the same time, summarizes the knowledge gained in his many years of activity as a physician. His occasional criticism of the authorities of previous epochs provides evidence of independent judgment on his part. The seven-volume handbook written by Paul of Aegina was understood by its author to be a presentation of practical medicine based on the medical literature of the past. He deals with all illnesses from the point of view of diagnosis, etiology and therapy. Paul explicitly testifies to his literary dependence on Oribasius.

The fully or partially extant early Byzantine commentaries on certain works by Hippocrates and Galen constitute a unique sort of literary evidence, in that they consist of transcripts of oral discourses. They document teaching in the medical training centres of the Byzantine age. Among the academic instructors active in 6th and 7th century Alexandria were Stephanus of Athens and John of Alexandria. Explanations of the Hippocratic writings "Prognosticon" and "Aphorisms" and the first book of Galen's "On Therapeutic Methods, Dedicated to Glaucon" are based on Stephanus' lectures, whereas fragments of the commentary to book VI of the Hippocratic "Epidemics" and a large fragment of the beginning of an interpretation of the Hippocratic text "On the Nature of Children" are based on the lectures of John.

In areas where Latin was spoken, medical literature in late antiquity was more practically oriented than in the eastern part of the empire. These areas also exhibited a tendency to introduce superstitious practices and cures from the so-called *Dreckapotheke* alongside scientific medical procedures:

> For *varuli*, i.e., styes on the eyes, prepare the following remedy: Remove the rings from your fingers, hold them with three fingers of your left hand around the eye, spit three times, and say three times: RICA, RICA, SORO. If a stye has formed on the right eye, hold it with three fingers of your left hand while looking east under the open sky, and say:
>
> The she-mule does not give birth, and the stone does not bear wool.
>
> This disease shall not grow a head, or if it grows, it shall disappear.
>
> After saying these words, touch the ground with the same three fingers, spit, and do this three times (*Magic spell from Marcellus, "On Remedies"*, 8,190f.: CML V, p. 161).

The collection of prescriptions entitled "On Remedies", which Marcellus of Bordeaux (c. 400), a high-ranking government official, composed as a layman for other laymen, is based primarily on older sources. Due to its author's origins, traces of Celtic language have been preserved in the collection. Dating from around the same period, a three-book treatise by Theodorus Priscianus entitled "Easily Obtained Remedies" described methods for treating external, internal and gynaecological illnesses. Cassius Felix followed Greek models in compiling his mid-5th-century work "On Medicine". The author's intention was that the work should contain everything that had proven appropriate for the treatment of the human body.

5.3. References

(Almost) complete editions of the works of Hippocrates and Galen are still, respectively, those of E. Littré (1839–1861) and C.G. Kühn (1821–1833).

Modern editions of works on ancient medicine are regularly published in the collections of the Corpus Medicorum Graecorum et Latinorum, Berlin, in the Collection des Universités de France, Les Belles Lettres, Paris and Loeb Classical Library, London/Cambridge, MA.

Many editions have appeared in other series; among these, an important one is Garofalo, I. (1991). *Galeno: Procedimenti anatomici*. Rizzoli, Milan.

Annotated editions of individual works or collections of fragments, including those of Diocles of Carystus, have been published in the collection *Studies in Ancient Medicine* from Brill (now De Gruyer-Brill), which also hosts many important studies.

Good, annotated translations of Galen's work are available in the *Cambridge Galen Translations* series (Cambridge University Press).

Projects of online encyclopedias with special sections on ancient medicine have been started, including the *Greek and Roman Humanities Encyclopedia* by De Gruyter-Brill.

Further bibliography is available in the works cited below:

Boudon-Millot, V. (2007). *Galien : Introduction générale. Sur l'ordre de ses propres livres, Sur ses propres livres. Que l'excellent médecin est aussi philosophe*. Les Belles Lettres, Paris.

Boudon-Millot, V. (2012). *Galien de Pergame. Un médecin grec à Rome*. Les Belles Lettres, Paris.

Bouras-Vallianatos, P. and Zipser, B. (2019). *Brill's Companion to the Reception of Galen*. Brill, Leiden.

Craik, E.M. (2015). *The Hippocratic Corpus: Content and Context*. Routledge, London/ New York.

Debru, A. (1996). *Le corps respirant. La pensée physiologique chez Galien*. Brill, Leiden.

Deichgräber, K. (1965). *Die griechische Empirikerschule*. Sammlung der Fragmente und Darstellung der Lehre, Weidmann, Berlin.

Durling, R.J. (1993). *A Dictionary of Medical Terms in Galen*. Brill, Leiden.

Edelstein, L. (1967). *Ancient Medicine*. O. Temkin, C.L. Temkin, Johns Hopkins University Press, Baltimore.

Edelstein, E.J. and Edelstein, L. (1988). *Asclepius: A Collection and Interpretation of the Testimonies*. Arno Press, New York.

van der Eijk, P.J. (1999). *Ancient Histories of Medicine*. Brill, Leiden.

van der Eijk, P.J. (2005). *Medicine and Philosophy in Classical Antiquity: Doctors and Philosophers on Nature, Soul, Health and Disease*. Cambridge University Press, Cambridge.

Hankinson, R.J. (2008). *The Cambridge Companion to Galen*. Cambridge University Press, Cambridge.

Jouanna, J. (1974). *Pour une archéologie de l'école de Cnide*. Les Belles Lettres, Paris.

Jouanna, J. (2012). *Greek Medicine from Hippocrates to Galen: Selected Papers*. Brill, Leiden.

Jouanna, J. (2017). *Hippocrate*. Les Belles Lettres, Paris (English translation *Hippocrates,* Johns Hopkins University Press, Baltimore, MD/London, 2001).

Jouanna, J. (2024). *Médecine et tragédie en Grèce antique. Scripta minora 1961-2023*. Les Belles Lettres, Paris.

Lewis, O. (2017). *Praxagoras of Cos on Arteries, Pulse and Pneuma,* Brill, Leiden.

Lloyd, G.E.R. (1991). *Methods and Problems in Greek Science: Selected Papers*. Cambridge University Press, Cambridge.

Lloyd, G.E.R. (2003). *In the Grip of Disease. Studies in the Greek Imagination*. Oxford University Press, Oxford.

Lloyd, G.E.R. and Sivin, N. (2002). *The Way and the Word. Science and Medicine in Early China and Greece*. Yale University Press, New Haven.

Manetti, D., and Roselli, A. (1994). Galeno commentatore di Ippocrate. *Aufstieg und Niedergang der Römischen Welt*, 37(2), 1529–1635.

Nutton, V. (2020). *Galen: A Thinking Doctor in Imperial Rome*. Routledge, London.

Nutton, V. (2024). *Ancient Medicine*, 3rd edition. Routledge, London.

Perilli, L. (2004). *Menodoto di Nicomedia. Contributo a una storia galeniana della medicina empirica*. Saur, Munich.

Pormann, P. (2018). *The Cambridge Companion to Hippocrates*. Cambridge University Press, Cambridge.

Singer, P.N. and Rosen, R. (2024). *The Oxford Handbook of Galen*. Oxford University Press, Oxford.

von Staden, H. (1989). *Herophilus: The Art of Medicine in Early Alexandria*. Cambridge University Press, Cambridge.

6

Medicine in Ancient Greece and Rome: Social Structures and Institutions

Vivian NUTTON
University College London, UK

The history of ancient Greek and Roman medicine begins and ends in darkness. The progressive impoverishment of the Western, Latin, Roman Empire following the barbarian invasions and its political fragmentation destroyed many of the institutions that had supported earlier medicine. What had been a flourishing and varied medical scene in the early fourth century (and until the mid-sixth century in Northern Africa) seems to have disappeared almost entirely by 600 CE, with the church the major source of learned healing. A greater continuity can be traced in the Eastern, Greek, half of the Empire, always far wealthier than its Western counterpart, but few texts survive that were composed in the late sixth or seventh centuries. By then Galenism had eclipsed its rivals and from the 550s, at the latest, had also become in translation the dominant form of learned medicine among the Christians of the Near East. In both East and West there is then a hiatus in our evidence.

6.1. Pre-classical Greece

The Homeric poems provide the first substantial information on medicine in the world of Archaic Greece. The *Iliad* begins with a medical crisis, the plague affecting

History of Life Sciences and Medicine 1,
coordinated by Lorenzo PERILLI.

the Greek army encamped before Troy, and with the initial determination by a religious seer of whether it was to be cured by religious means or by the intervention of a doctor. The fact that it was decided that its cause lay in the anger of Apollo at the treatment by Achilles of the daughter of one of his priests, and that it required therefore a religious response from Achilles and the Greeks, shows only that the two types of explanation coexisted together. Whenever the poem as we have it was composed, probably between 750 and 700 BCE, in the eyes of the poet and his audience at the time both possibilities were to be considered. Certainly, the author could employ an already detailed, varied and often precise vocabulary when describing the nature of the many wounds among the warriors, and he was aware of the reputation of Egypt for medical herbs. Some of his doctors, notably Asclepius, Machaon and Podalirius, lead their own military contingents, and other anonymous doctors are also mentioned, evidence perhaps for wide status differences between individual healers.

A further clue to the role of doctors at this time and earlier is found in the Homeric association of bard, doctor and armorer. All three attend upon noble patrons, and travel from one palace to another to provide specific services. This pattern may be found much earlier in the presence of a physician among the tables recovered from the palace at Pylos (S. Greece) around 1200 BCE. There is even earlier evidence for physicians being sent from one monarch to another as part of the diplomacy of Near Eastern rulers. The mention of an Achaean healing deity among those invoked to cure a Hittite ruler and the possible sending of a statue of the god also provide a context for these wandering professionals, many if not most of whom would have learned their skills within their families. That this pattern continued for centuries is neatly indicated by a bronze tablet from Idalion, a small kingdom on Cyprus around 476 BCE. When the Persians and the inhabitants of Citium were laying siege to the town, King Stasikypros and his subjects invited the doctor Onasilas and his brothers to come and treat the wounded. Their reward for their free treatment was two parcels of land, along with their crops: if they were later driven off this land, the perpetrator should pay a monetary sum equal to the value of the land stated in the agreement. There are other examples of Greek doctors moving to treat the King of Persia, notably the historian Ctesias of Cnidus (around 400 BCE) and, a century earlier, Democedes of Croton, whose career took him from Southern Italy via the island of Aegina and Athens to the court of King Darius. The legend that the great Hippocrates turned down a substantial reward to attend the King of Persia and refused to treat this enemy of Greece, a fabrication of the Hellenistic period, sheds light on both court doctors and the ethics of medical dealings with outsiders. Later writers praise Hippocrates' patriotism and are silent on any contradiction of the universal benevolence insisted upon in the *Oath*.

6.2. Greek medicine and the Ancient Near East

The presence of Greek doctors in the Near East and the transfer of healing substances in archaic Greece from there to the Greek world raises a major question that modern scholarship has still to resolve adequately: the relationship between the medicine of Greece, Egypt and Babylonia. Some points are clear. Babylonian medicine was alive and well in the Levant in the fourth century BCE, and probably centuries later in some form of Talmudic medicine among the Jews. Egyptian medicine continued to flourish alongside Greek medicine well into the Roman period, when libraries at its temples could display writings in both traditions. Egyptian writings mention Cretan beans and other remedies from the Greek world, while aromatic resins and such substances as opium and Cyperus mentioned in Greek texts had a Near Eastern origin. The gynecological texts in the Hippocratic Corpus contain many such ingredients, and the practice of fumigation, applying sweet-smelling vapors and scents to the genital area, may also have come from there. One writer from Egypt, Ninyas, appears on the list of those who ascribed illness to residues of food remaining in the body, an idea also found in early Egyptian texts. One can also point to similarities with Babylonian medicine and medical practices, which the historian Herodotus found little difficulty in describing in Greek terms. The Babylonian emphasis on fluids and balance and their, at least partial, distinction between religious and more natural cures has also suggested to scholars, Mark Geller in particular, that Babylonian doctors exercised a more powerful influence on early Greek medical thought than their Egyptian counterparts.

6.3. Classical Greece

But while such an interchange of ideas may have taken place, what survives of Greek medicine before 400 BCE shows a very different social and intellectual picture. While clans of doctors survive, most notably those, like Hippocrates, who claimed descent over generations from doctors, and even from healing deities like Asclepius, Apollo and Hercules, medicine was by then not a closed profession. The so-called Hippocratic *Oath* implies a move away from a clan system toward one that can be termed an apprenticeship system, something regarded as normal by Plato writing in the 370s BCE. The discussions around 500 BCE involving medicine and health, including theories of the body, were often carried on by philosophers or sages, many of whom were not doctors themselves but were interested in the human body as an example of the workings of the natural order of creation. Far from there being any tenets that all must obey, an obvious characteristic of these earlier writings is their combative and often public nature, as writers sought to defend their own individual point of view with great vigor and no little rhetoric. While some of these writings, not least *Aphorisms*, *Coan Prognoses* and *Dentition*, contain memorable statements, others are unintelligible except to their author or require a teacher to explain them properly to the beginner. Others, such as the *Art*, are public performances aimed at a non-professional audience of potential patients. Still others, like *Ancient Medicine* or *Sacred Disease*, defend the author's ideas in public and attempt to widen the scope for non-religious explanations of disease. This combative, public approach is not found in Egypt or Babylonia, and may be related to the wider public debates on politics that characterize both the Greek democracies and, to judge from Homer and the discussions described by the historians Herodotus, Thucydides and Xenophon, the more restricted Greek aristocracies. It is a characteristic that, as we shall see, continues into at least the third century CE, and indicates the relative openness of the ancient Greek (and later Roman) face-to-face society.

Information about medicine and medical practice is scanty before the fifth century BCE, with a few scattered quotations, occasional representations in art, and passing references in poetry, but no surviving medical text. By contrast, the period from 470 to 320 BCE offers an enormous variety of written material, both technical and non-technical, as well as many inscriptions recording the names and even careers of doctors. The Greek world had also expanded in the previous two centuries from the Aegean basin to include colonies on the coast of modern Libya and, with greater wealth and power, on the sea coasts of Southern Italy as far North as Naples, most of Sicily, and even on the Black Sea. These regions were in frequent trading and cultural contact with the world of the Aegean, and doctors were among the frequent travelers. Some tracts in the so-called Hippocratic Corpus may well have been written by a Western Greek, for well-known medical centers such as Croton (Crotone) were to be found there. Democedes moved from Croton to the island of Aegina, and then crossed the bay to Athens, before leaving to join the court of Persia. The author of *Epidemics I* and *III* seems to have moved from somewhere in the Southern Aegean to practice along with other doctors on the island of Thasos on the northern coast of the Aegean and may also have visited the mainland. The travels of the writer of *Airs, Waters and Places* took him even further, for he described the ailments and constitutions of the native Libyans, the inhabitants of modern Western Turkey, and the Scythians, the nomads of the steppe to the north of the Black Sea. He wrote his treatise in order to help doctors like himself, who might be called on to treat patients from cities and regions with which they were unfamiliar and who needed a brief introduction to the relationship between individual illness and the environment.

But it would be wrong to imagine all ancient Greek healers constantly on the move in search of patients, for there is plenty of evidence for many remaining for long periods in a large city, such as Athens or Corinth, and using tools and substances that were not easy to transport around. The large so-called Hippocratic bench, a very large block of wood that was used by the practitioner in treating dislocations and sprains, would have been almost impossible to load onto a horse or mule. Nor is it easy to see how that author of *Epidemics I* and *III* could have compiled his descriptions of the courses of illnesses over several years from his own and his colleagues' notes without there being a central location where they could be stored and consulted. Even in rural districts with a sparser and more scattered population, there will have been healers who were largely resident in one place, but who embarked at regular intervals on visits to patients in the countryside. Such a model gains added weight from later evidence for such practices in Roman Gaul, where towns like Bar-le-Duc appear to be a center from which healers traveled out, and in late-Antique Cilicia, where the Christian healers Cosmas and Damian were imagined as taking it in turns to go out for several days from their base to treat both humans and animals. One of the later Greek words for such a doctor, *periodeutes*, a man on a circuit, implies that such a circuit was common. Recent work on so-called charlatans in Renaissance Italy has insisted strongly that such peripatetic practitioners were a valued, and even necessary, part of the healing spectrum, particularly in areas that lacked major urban centers.

One way in which communities could ensure a permanent presence of a doctor was to make a special payment to an individual along the lines already noted in the Idalion contract. If the historian Herodotus is to be believed, this was already happening in the late sixth century for both Aegina and Athens were said to have paid Democedes for his services. But there is then a gap in the historical record for a century, and the most detailed and widespread evidence does not appear until Hellenistic times. At Athens, certain physicians were chosen by the assembly to be on hand and to treat its citizens in the same way as they were selected to serve on campaign or in the fleet. Whether this 'retaining fee' compelled them to offer their services to citizens for free is a vexed question, but certainly slaves were excluded, left to be attended by less able healers, and possibly even fellow slaves. Later documents show doctors being praised for not accepting money, but this is more likely to have been their own decision rather than a legal obligation. Social pressures in a small community might impose free treatment for the poorer members of the community, but this would not have been the case for rich citizens or for non-citizens. Nor is there any need to assume that a public doctor had at this date the contractual duty of assisting at inquests or on other official occasions (situations known from much later documents from Roman Egypt), or that only he was allowed to speak as an expert witness in court. Medieval and renaissance contracts for town doctors show a great variety of obligations, as well as the gap that might arise between reality and legal requirements, and there is reason to assume a similar situation in Antiquity.

6.4. Literacy and a wider interest in medicine

The medical writings that form the so-called Hippocratic Corpus, a term that is far from precise, also mark a major development in ancient Greece, the growth of literacy. Some of the texts within it are among the earliest surviving prose texts, let alone scientific texts, in European medicine, and display a substantial level of competence in composition, and literary techniques that go far beyond the production of case-notes or the listing of recipes. Many refer to other writings, occasionally with a name, as part of the author's attempt to establish the superiority of his own ideas. The author of the section in the Anonymous London papyrus dealing with competing medical theories, often associated with Meno, a pupil of Aristotle around 330 BCE, was able to associate many individuals, not all of whom would qualify as medical practitioners, the philosopher Plato among them, with specific ideas, which suggests that by then such summaries were easily available and would have ranged far wider than what is available to modern scholars.

The advent of literacy only strengthened the move away from a clan-basis of medical practice toward an open art, accessible in many ways to those who wished to gain medical information and knowledge. They could hear public performances, some of them later in the Hellenistic period sponsored, or at least approved of, by a town council. Others paid to be taught: Plato regarded Hippocrates of Cos as a great teacher, the equivalent of the sophist Gorgias in rhetorical teaching. Still others shared with doctors in after-dinner discussions of medicine and medical theory. Plato's *Symposium*, for example, includes a doctor, Eryximachus, among his interlocutors, thus providing a precedent for such later writers as the Roman Plutarch or Methodius of Patara, a Christian writer, to introduce a medical perspective on wider themes. Others would have picked up their knowledge simply from reading books and used it for their own purposes. The historian Thucydides, writing in the last decades of the fifth century, not only used medical terms to describe the effects of the plague in Athens and elsewhere, but his analysis of the causes of the Peloponnesian war resembles the arguments of doctors in debating the causes of disease. Comedians in the late fifth century made jokes about the incompetence and verbosity of physicians (similar jokes about pharmacists seem to have emerged a generation or so later) and set a precedent for centuries to come. Dramatists like Sophocles in his *Oedipus Rex* and *Philoctetes* presented medical dilemmas on stage to an appreciative audience. The philosopher Plato knew some of the doctrines of Hippocrates, and praised him for his acute logic, although not so precisely as to enable modern scholars to identify any specific tract in our Hippocratic Corpus. His knowledge of medical debates is most clearly shown in his *Timaeus,* an influential account of creation that gained for him the longest entry in the anonymous account of medical ideas already mentioned. In addition to the widespread belief in balance and imbalance, as well as the failure of air to move properly around the body, he posited that the healthy process by which the elements combined to produce all the constituents of the body could go into reverse, flesh could decompose and, in the most serious illness, the very marrow of the bones could be destroyed. However much doctors might regret the incursion of non-medics into their territory, this was merely one feature of the open nature of Greek medicine from at least the fifth century BCE until Late Antiquity. It was this participation of a variety of individuals that contributed in this period to what has been termed the emergence of the physical body in Ancient Greece, as physicians and laymen, not to mention artists and sculptors, brought their own insights to bear on what it was to be a human being.

6.5. Religion and medicine

This emphasis on investigating and describing how the body worked in both sickness and health still left a place for the gods in the healing process. The author of *Sacred Disease*, while attacking the involvement of soothsayers and religious healers in dealing with epilepsy, does not attack the gods but reinterprets the role of the divine. He rejects prayers and incantations as impious attempts to force the divinity to do what the suppliant demands, and instead stresses the power of the gods in their work of creation that allows healing to take place. Another writer declares that prayer is a good thing, but, at the same time, the human healer should be prepared to lend a hand. It was a sentiment shared by most doctors. By 350 BCE the temple of Asclepius at Athens served as a meeting place for doctors, who set up their memorials and even their instruments on its walls in acknowledgment of the power of the god. A beautiful relief from the Asclepieion from the mid-fourth century shows five men in prayer before Asclepius, Demeter and Kore (the latter pair strongly associated with the Eleusinian mysteries). One man has been identified with a famous physician, Mnesitheus, and two more are sons of yet another, Dieuches. Honorary decrees from at least the late fourth century were set up there and in other temples of Asclepius, most notably at Cos, praising doctors who had performed distinguished state services.

Such healing shrines were ubiquitous, many of them in Attica to local heroes such as the Hero Healer Dexion, others, such as the cult of the Thracian rider, brought from outside by the many foreigners who flocked to a prosperous Athens.

The most famous of all immigrant healing divinities was Asclepius, whose cult seems to have originated in Thessaly, although some believed that it was begun in the Peloponnese at Epidaurus. There was also a long tradition linking the god with healers from the island of Cos. What is clear from the archaeological evidence is that before 430 BCE this regional cult, and that its spread around the Aegean basin owed much to the influence of Athens among its imperial subjects. There is also likely to be some link with the most deadly of epidemics of Athens, the Athenian plague of 430, an as yet unidentified disease that devastated the population shut up within the city's walls in response to a Spartan invasion. Thucydides, himself a sufferer, claims that it had traveled from Egypt to Athens and its troops at Potidaea in Northern Greece, but also attacking 'many populous places', although not in the Peloponnese itself. The Roman historian Livy, writing many centuries later, mentions epidemics of pestilence in Rome in 431 and 428 BCE, although without making any connection between the outbreaks (a typical feature of ancient reporting of epidemics, which are heavily focused on the local). Thucydides' account, which became the model for many later historians, concentrates on the social consequences of an epidemic that none could conquer: the doctors did what they could, but to no avail, nor could succor be obtained from the gods. His picture of a society disintegrating in the face of disease, a contrast with the optimistic speech of the Athenian leader Pericles only a few months early, gives the lie to the later story that it was none other than Hippocrates himself who had banished the plague by burning bonfires in the streets and purifying the air (a legend that was widely believed and may have influenced Health Officials in Renaissance Italy to follow his example).

What is striking is the relative speed and the extent to which the cult of Asclepius spread around Greece from the 420s, subsuming or overshadowing local cults, with Asclepius often taking over from his father, Apollo, as at Epidaurus itself, the earliest temple seems to have been dedicated to Apollo Maleatas. When later Asclepius made his way from there to Athens, the god stopped briefly at Eleusis to be initiated into the Eleusinian mysteries, where a festival called the Epidauria seems to have been established by around 400 BCE. Only in Boeotia, where a local healer, Amphiaraus, had a similar incubation shrine near Oropus, did the Asclepius cult fail to gain a strong hold at first. Asclepius was frequently seen as an outsider. The story of the introduction of his cult into Athens represents it as an originally private decision (although one that required approval from the religious authorities), as the god moved from the island of Aegina first to the port of Piraeus around 423/2 BCE and finally to the slopes of the Acropolis two or three years later, where he was joined by his daughter Hygieia soon after. The shrine itself, which included a small grove, was not completed for five or six years, but was quickly supported by many leading citizens, including the dramatist Sophocles. In

399 BCE, Socrates, in his deathbed, reminded his friends that he still owed a cock to Asclepius, although what he meant by this remains obscure.

The position of the Athenian shrine of Asclepius, just outside the main religious precinct of the city, is typical of his temples: the striking exception, that of Messene, may be explained by the tradition that Asclepius was a fellow townsman. Very often, as at Epidaurus, Cos and, later, Pergamum, the shrine itself was located some distance away from the town, prompting one later author, Plutarch, to suggest that this might have been done for reasons of health. Certainly, the fact that one had to make a pilgrimage some way out of the town may have added to the belief of the suppliant that the healing process was something special. Most shrines were small, but others, most noticeably during this period, Epidaurus, were capable of receiving thousands of pilgrims. As the most famous center of the cult, Epidaurus expanded enormously in the fourth century. The ivory and gold cult image, commissioned from a famous sculptor, was huge, albeit half the size of that of Zeus at Olympia, and its buildings were painted by celebrated artists. Its theater, built around 350 BCE, was regarded as the most elegant in the whole world, and it was the site of musical and dramatic contests in honor of Asclepius. Suppliants would come to worship at the shrine, and spend a night in special incubation rooms, where they might receive a vision from the god that would then be interpreted by priestly officials or would directly result in a cure.

Some of the results of these divine ministrations were further engraved in stone around the walls of the shrine for others to read of the power of the god, in punishing those who doubted him as well as repairing a broken goblet, finding a lost child and curing a variety of conditions from paralysis and facial blemishes to problems with pregnancy – Asclepius seems to have had a monopoly on three- and five-year pregnancies. These accounts, together with masses of votive offerings from shrines such as that at Corinth, suggest that many sufferers came with chronic problems, but others may have chosen to come first to the shrine with any ailment or, perhaps more typically, to have tried both a religious and a non-religious cure.

But although Asclepius became the most famous of all healing gods, he was far from alone. The author of *Regimen* mentioned Zeus, Athena, Apollo, Hermes, Earth, Sun and the Heroes as gods who might be invoked to avert disease, and an oracle of Apollo from Delphi to the Athenians in 348 BCE ordered them to sacrifice for the sake of their health to Zeus the Most High, Hercules and Apollo the Protector. Girls and young women made their healing vows at Brauron in central Attica to Apollo's sister Artemis, whose healing activities were sometimes shared with Eilythuia, the goddess of childbirth.

6.6. Human healers

This variety of healing gods is mirrored in the variety of human healers. Some, like wandering shamans, might employ chants, charms and purifications; others, the so-called 'rootcutters', offered herbal remedies, while drug sellers preferred more complex or exotic remedies. Some places, most notably Cos, were famous for their healers, and towns around the Aegean later sent to the island for suitable candidates to be a civic doctor. How a doctor might progress is neatly illustrated by an honorary decree from Halasarna on Cos. Onasander had come with his master Antipater from Halasarna to the main town of Cos, eventually setting up his own practice and treating both Coans and individuals from his Halasarna. Some of them he treated for nothing, even if they were very ill and might need long-term care, and his kindness gained him official honors, like those given earlier to Antipater.

The role of women both as patients and healers is far from clear. Certainly, there was no bar to their treatment by male doctors, although, as the nurse in Euripides' play *Hippolytus* explained, there were some intimate problems that would be discussed only with fellow women. Male doctors also attended births, but Plato's comment that only elderly women, mothers themselves and beyond childbearing age, acted as midwives, *maiai*, is exaggerated. At least one woman, Phanostrate, extended her treatments beyond the obstetric and called herself a 'doctor'. Presumably, she also treated children, and it cannot be excluded that she also treated some men, although social pressures are likely to have kept their numbers low. But it was expected that the mistress of the house would be able to supervise the health of slaves, and she may have been the first person to be called upon to treat minor injuries or ailments in the family. Given the links between cooking and the preparation of remedies, a *common* analogy in medical texts, women's expertise may not have been far removed from that of many male doctors and allowed some to sell their remedies in public.

This plurality of healers was assisted by the absence of any official standard or qualification to practice such as those developed in the Middle Ages and later. Family connections would help to suggest, although not guarantee, competence. Doctors from Cos had a great reputation, but, as yet, there is no clear evidence for others traveling there to gain instruction. Long residence in a town would help, but the problem of initial credibility affected both patient and practitioner. Setting up practice in a new locality was not easy at first, and entrusting oneself to the hands of an unknown doctor whom one had never seen before was an act of faith, if not of desperation.

Medical writers of the fifth and fourth centuries BCE offer two assurances to potential patients. One, expressed in such public documents as the *Art* or the *Law*, was to emphasize the moral character of the practitioner and his superior expertise. The most famous, and most disputed, of all medical writings, the so-called *Hippocratic Oath* (which few scholars today believe has any link with the historical Hippocrates), sets out guidance for those from non-medical backgrounds who have decided to learn medicine. It emphasizes the student's duty to his teacher and the teacher's family, as well as to his patients and their household. Two clauses, contentious even today, forbid the administration of deadly substances and the giving of some form of abortive pessary, although both euthanasia and abortion seem to have been widely practiced in Antiquity. When the *Oath* was composed and under what circumstances are still disputed, but it is clear that the document itself had become well known by the early second century BCE, when it or, less likely, a similar *Oath* was held up by the Roman politician Cato as a prime example of the way in which Greek doctors banded together to harm their patients. (Other similar interpretations of the *Oath* as a conspiracy against patients can be found from the Renaissance onward.).

The second element in a doctor's armory, emphasized above all by Galen as the essential element in what he saw as the medical practice of the great Hippocrates, was the art of prognosis. More than is implied by the modern word, it involved, according to the author of *Prognostic,* a knowledge of the past, present and future of the patient (although how far back in the recording of an illness a doctor would go is uncertain) in his or her environment in order to decide on the future course and treatment of the illness. While this would enable the doctor to plan appropriate action, it also allowed the patient to make a judgment on his ability. If the case turned out as he had predicted, that was proof of his competence and would also allow the patient and his family to place their trust further in him. This author emphasized, rightly, the importance of trust in assisting the healing process. But it can also act as an insurance policy. Were the doctor to decide that the case was incurable, and also reveal this to the patient's family, he would avoid any responsibility for the death of the patient and would not be criticized if he withdrew from the case or handed it over to someone more experienced, for example, in surgery. (Most doctors would appear to have had a basic grounding in surgery, but there were others, particularly if they had served as military surgeons, who were acknowledged to have had much greater experience.) The decision not to inflict unnecessary harm on a patient was also approved by all sides. But if, after predicting the worst, a practitioner offered to try and succeeded in effecting a cure, this was an added indication of his abilities. Prognosis, as defined in this treatise, as well as in such tracts as *Airs, Waters* and *Places,* was thus both a social and a medical necessity.

Medicine in the first sixty years of the fourth century BCE was only tangentially affected by the confused political situation in the Aegean world. The first Athenian Empire had been lost after the defeat in the Peloponnesian War in 404, and a later attempt to re-establish it also proved short-lived. Other city states that tried to establish dominance, including first Sparta and then Thebes, failed likewise. Major powers on the fringes of the Greek world pressed hard on smaller Greek cities. In the West, the Carthaginians were a constant threat to the Greeks on Sicily, while in the North, Philip II, King of Macedon, gradually came to extend his hegemony over mainland Greece. The King of Persia, whose attempts at conquest had been triumphantly defeated in the wars with the Greeks and the Athenian League, now recovered control over most of the wealthy Greek cities of the Anatolian coast. Memories of their perhaps distant independence may have generated some of the legends about Hippocrates, who was said to have refused to go to Persia and attend its ruler because he was an enemy of the Greeks, hardly a recommendation for a doctor's duty to all in need. Another story told of his cure of Perdiccas, King of Macedon, who was suffering from love-sickness, only one, and perhaps not the earliest, of a series of stories involving this troublesome condition. A later monarch, Philip II, was attended by, among others, the father of the distinguished philosopher-scientist, Aristotle, who taught the future Alexander the Great and who later migrated under Macedonian rule to Athens, where he set up his own 'school', the Lyceum, as a rival to the Academy of his own teacher, Plato. Both groups debated questions about the workings of the human body. Aristotle and his contemporary Diocles of Carystus were interested in animal dissections, and their findings were discussed by other philosophical groups, most notably the Stoics. Plato, Aristotle and the Stoics were particularly interested in teleology, a way of explaining the body in terms of the purposes and functions it performed in accordance with the plan of the divine Creator, Nature or the Demiurge. These philosophical disputations, particularly on the workings of the soul, imparted a certain rigidity to these intellectual groupings that had not been there before and that became mirrored in the development of the so-called medical sects. Although both 'logical' or 'dogmatic' practitioners and their Empiricist opponents looked back to Hippocrates and even earlier, it would be wrong to push back into the third, let alone the fourth, century BCE all the doctrinal positions that are recorded from much later centuries. Philosophers disputed, in part from medical evidence, whether the human soul was a unity located in the heart, as Aristotle had argued, or, following Plato, if it consisted of three interconnected parts, each in its own seat, the brain, the heart and the abdomen or liver. Similar debates within medicine concerned mainly whether one should use logic following on a long, complex and detailed observation of the individual patient to reach a conclusion as to what was the cause of an ailment and, once that had been identified, to devise an appropriate therapy to remove it. Or was it easier, quicker and more reliable to discover similar cases from the past and apply whatever had worked then to the case under discussion? Both sides appealed to logic and learning but in different ways, and both were agreed that diagnosis and treatment were far from simple.

6.7. The Hellenistic world

The conquests of Alexander the Great, d. 323 BCE, and, although less noticed at the time, the expansion of Rome altered fundamentally the political structures and geographical horizons of the Greek world. Alexander's army defeated the Achaemenid rulers of Persia and took over the Middle East from modern Western Turkey to Afghanistan and the Punjab. Settlements of Greeks even in remote areas imposed their culture on their new subjects, and even when control was relatively swiftly lost, particularly in the East, the impetus to adopt Greek ways in art and culture long remained. It is not clear how far one can trust later reports that, rather like Napoleon in Egypt, Alexander had arranged for scientists of various kinds to accompany his troops and bring back detailed information, and even specimens, to Greece. But the contacts with the newly conquered lands gave access to previously unknown or unused plants and minerals, which tripled the number of medicinal species known. The difference between the materia medica in the Hippocratic Corpus and what is recorded by writers from the first century CE is substantial, although one cannot give precise dates for their introduction.

Crucial to this development was the foundation of the city of Alexandria 'by Egypt' in 331 BCE. Sited on an island just off the coast near one of the mouths of the Nile, it both commanded the waterway into the interior and was easily defensible against aggression from land and sea. It gave access to the great granaries of the Nile valley, and quickly became the largest city in the ancient world as well as its most significant entrepot. Goods, including drugs, could reach it overland across the desert from modern Iraq and Iran, via the Nile from the Sudan and East Africa, or by boat from India and further afield to its Red Sea ports, where they were then transhipped to the Nile. As the former gateway to Egypt, ships could sail there from all over the Mediterranean to exchange their cargo. At first only Greeks were allowed to settle there with full rights, but over time, the city became an international metropolis with major concentrations of Jews as well as of native Egyptians.

Alexander and his father, Philip II, had already sponsored a court culture in their capitals, and their example was quickly followed by Alexander's generals when they carved out kingdoms for themselves after his sudden death. Although the Macedonian kings still retained control over most of mainland Greece, they were challenged by warring generals, particularly Seleucus and his successors, who ruled over much of the Middle East from their capital at Antioch (Antakya, Turkey), and by Ptolemy I, who made Alexandria his capital. He and his immediate successors set out to make the city a showpiece of Greek culture, whether in its buildings, including the magnificent lighthouse, or in the intellectuals who flocked to its court.

Two institutions together shifted the intellectual center of the Greek world away from Greece (beset by conflicts between the competing generals) toward Alexandria and served as a model for future capitals. Most is known about the Alexandrian library or libraries, which were deliberately created to house all previous Greek literature under the eye of intellectual and scholarly librarians. According to Galen, Ptolemy confiscated for his library any book that had not been previously recorded, supplying a copy in return. These included many medical texts with Coan connections, for the island itself long formed part of the Ptolemaic realm, and it is probable, although far from certain, that this is where what we call the Hippocratic Corpus was formed as books by different authors, many anonymous, were juxtaposed on the shelves of the Great Library.

Equally significant was the Museum, the Hall of the Muses, a temple complex where poets, historians, scientists and medical men could congregate and discuss intellectual topics at the expense of the ruler. Some like the polymath Eratosthenes or the poet-librarian Callimachus were provided with lucrative posts, or at least if one can rely on the title of members in the Roman period, they were provided with free meals in the Museum, rather like Oxbridge dons. Alexandria became the intellectual center of the new Greek (or Greekish, the meaning of Hellenistic) world for at least two centuries and the model for both Antioch and, in the second century BCE, Pergamum. The rulers of the latter built a magnificent new capital with, at the summit of the great acropolis, an enormous library and an equally massive altar, now in Berlin, that commemorated their triumph over invading Gauls in the middle of the third century BCE.

Doctors as well as literary figures were attracted to the new city of Alexandria in hope of official patronage or because of the opportunities for innovation that were on offer in the rapidly expanding city. One of those who arrived in the early third century BCE was a man from Chalcedon, a small town on the Bosphorus, Herophilus. He is said to have studied with a distinguished doctor from a medical family on Cos, Praxagoras, who may well have been responsible for his interest in some of the writings associated with Hippocrates and also in anatomical dissection. Systematic dissection, as opposed to occasional observations made during surgery or medical (or veterinary) practice, cannot be documented in Greece before the middle of the fourth century BCE. Both Diocles of Carystus and Aristotle are known to have performed dissections on animals, although precise details of what Diocles did and of his relationship with Aristotle are lacking. However, the series of dissections made by Aristotle of a variety of species show that this was part of an organized program of research, whose methodology was transmitted to his pupils such as Theophrastus. Praxagoras pursued a similar interest in the decades immediately following the death of Aristotle in 322 BCE, investigating the blood vessels and correctly distinguishing between veins and arteries as two separate systems. He was the first to offer a detailed explanation for the workings of the pulse and its relationship to the heart, positing that it played a major role in facilitating the flow of *pneuma*, a form of refined air, around the body. This *pneuma* was transmitted through the vessels to *neura*, cord-like structures, that later anatomists identified with the nerves that assisted the body's movements, and almost certainly its sensory ability, although it would be going too far to credit him with the discovery of the nerves or with a sharp distinction between motor and sensory types. His anatomical investigations were also carried out within a clinical context and may not have resembled the more systematic investigations of his pupil, let alone those of Galen much later.

Diocles, Aristotle and Praxagoras were restricted in what they could dissect to non-humans. The Greeks had many taboos about touching a dead body, some of them officially promulgated in so-called Sacred Laws, and it would have been extremely difficult to get hold of a dead body. Earlier descriptions of the human body derive either from chance-isolated observation or from analogy with animal anatomy. But there was one place familiar to the Greeks where it was thought normal for organs to be removed from the human body after death, Egypt. Mummification was almost a tourist attraction for visitors and had been described by Herodotus and doubtless others before him. Appropriately trained personnel would first remove the contents of the brain through the nose of the corpse before opening the body and removing the lungs and most of the contents of the abdomen before placing them in four large Canopic jars: the heart and kidneys were usually kept within the body, which retained its skin, muscles and skeleton for mummification. How far the results of this expertise were transmitted to doctors is problematic, given the very small number of surviving Egyptian papyri. There may have been family links between the two groups, although the status of mummifiers ranked below that of the doctor, and the same Egyptian word could be used for both. Egyptian medical texts employ a large and varied vocabulary to describe the parts of the body, as well as a reasonably accurate osteology, and a Christian author, Clement of Alexandria, knew of an Egyptian book on anatomy. Nor can one exclude the possibility that some Greeks gained knowledge of some of the internal arrangements of the body from talking with native mummifiers or doctors, although linguistic difficulties would have hampered communication, but in general there is nothing to suggest that what was known about human anatomy by the Egyptians was superior to that of their Greek contemporaries in the fourth century BCE or had been transmitted to them. Rather, what was crucial was the apparent Egyptian acceptance of what to Greeks amounted to interference with a corpse, not least that of one's own kin. If this appeared to be a matter of indifference to the natives, there was no reason why Greeks should feel concerned about opening up a barbarian corpse.

The growing interest in dissection, allied to the possibility of working in a culture with ostensibly few taboos about dealing with a corpse, goes some way to explaining why human dissection is associated above all with Alexandria in the early decades of the third century. It is possible that Erasistratus dissected in Antioch, where his father, and later, he himself, was a court physician, but no ancient source mentions any place other than Alexandria. The Roman writer Celsus, around 35 CE, opens his account of medicine in his encyclopedia of *Artes* suitable for his gentleman reader (it included sections also on farming, rhetoric and generalship), with a history of medicine that describes in detail the Alexandrian debates about the origins and value of human anatomy. Celsus was relying on a Hellenistic Greek source that he interpreted for his Latin readers and that owed a great deal to the learned Alexandrian tradition.

Celsus makes it clear that human dissection was contested almost from the start and was made possible only by the agreement, and, one may suspect, the protection of the Ptolemaic ruler himself. As in other later stories from Antiquity of experiments on humans, the subjects had in some way forfeited their rights – in this instance as condemned criminals (and presumably non-citizen natives). That humans were dissected for research purposes is certain from what was found, but Celsus goes further in claiming that they were anatomized while they were still alive. It is possible that reports on the workings of the internal body are based on a combination of animal vivisection and post-mortem human dissection, but there is no a priori reason to doubt that such deadly experimentation could have taken place in the apartheid situation of early Alexandria.

The story of Alexandrian anatomy is largely focused on two individuals, Herophilus and the probably slightly younger Erasistratus of Ceos. To judge from the surviving fragments of their writings, it was Herophilus who first mapped the interior of the human body and named many of its parts, such as the duodenum and the torcular ('wine-press'), the slightly depressed junction of the skull sutures. He provided a careful description of the coats of the eye and the first detailed description of the liver and possibly the pancreas, as well as continuing his master's investigations into nerves and the pulse (an interest he transmitted to his pupils). He knew of some writings associated with Hippocrates, and believed in humors but not in the same way as the author of *Nature of Man*. He also took a more restricted view of prognosis in his practical medical advice; but, overall, he was viewed very positively by later writers, compared with his coeval Erasistratus.

Erasistratus, in part thanks to Galen's later criticisms, remains controversial even today. Full of ideas, many of them borrowed from contemporary debates with other scientists, he comes across as a very different personality from Herophilus. A brilliant anatomist, he made discoveries about the heart and brain that were not surpassed for centuries, as well as challenging many ideas about humors, which he viewed simply as fluids rather than the body's building blocks. He rejected Aristotelian and Platonic teleology: in his view, muscles were not formed ab initio to carry out a specific action but developed by carrying out that action. If Herophilus' body was largely a static one, that of Erasistratus was a moving and mechanical one. The stomach grinds and crushes food like a corn-mill, producing sensations of hunger when it continues to grind in the absence of food. The kidneys, liver and bladder act as filters, while his notion of an intertwining threefold rope of nerves, veins and arteries can be compared with discussions on torsion and rope-strength in artillery manuals. His conclusions were buttressed by experiment, whether examining pulsation and blood flow (where he utilized new ideas about a vacuum) or testing to see how much was lost through perspiration and transpiration. On one occasion he is said to have opened the abdomen so as to apply substances directly to the liver. This mechanical view of the body, and in particular his belief that the arteries carried only *pneuma*, earned him the disdain of Galen, although he was even more scornful about the followers of Erasistratus, who, he asserted, ignored or misinterpreted many of his theories.

This polarization between the two great anatomists should not disguise one important fact. With the exception of another dissector, Eudemus, whose dates and place of work are far from certain, we know of no other investigative anatomist for many centuries to come. Celsus regarded this type of dissection as long over, abandoned for many reasons: it was cruel, investigation into a dead body was misleading, for a living body was something very different from a corpse, and repeated dissections added little or nothing to knowledge of the body. But skeletal anatomy at least continued to be taught at Alexandria for centuries, even if it had ended elsewhere. A lack of interest on the part of the authorities and the expulsion of many intellectuals from Alexandria in the second century BCE, the withdrawal of royal protection, and the growing assimilation of Greeks and Egyptians will all have contributed to closing the briefly opened window of human dissection.

This does not mean that these discoveries did not have wider implications. Poets like Callimachus and Posidippus drew on this new information in their poetry, and the author of a pseudonymous Hippocratic tract on the heart, written perhaps around

250 BCE, seems to have known of Erasistratus' discovery of the heart valves. Further proof can be found in the rapid development of surgery in Alexandria, for Celsus records many more and more complex surgical operations based on a more accurate understanding of the structures of the human body than had been performed earlier. A rare surviving book from pre-Roman Egypt, by Apollonius of Citium in Cyprus, not only reveals new techniques for dealing with fractures and dislocations but also exploits anatomical information to explain the superiority of certain treatments for a dislocated hip.

Above all, the increasing prosperity of the world of the Eastern Mediterranean allowed cities to flourish and to imitate others larger than themselves. The favorable climate, ideal for agriculture, that lasted from the late sixth century BCE for a millennium or more, permitted the cities and towns of the Eastern Mediterranean to produce more food and become more prosperous, which in turn allowed intellectual institutions, including medical ones, to flourish on the profits of surpluses. There was also a deliberate attempt by non-Greeks in this period to claim a share in their Greek inheritance and to assert their Hellenic identity. This was not an even process, but Hellenization proceeded apace even in the face of widespread objections marked by revolts, as in Palestine. Earlier medical traditions remained, but increasingly it was Greek medicine that became dominant.

6.8. Medicine moves West

More significant even than the expansion of the Greek world to the East in this period were developments taking place in Italy, where the Latin-speaking inhabitants of Rome had established hegemony over the surrounding region of Latium and then, after a long series of wars, over the Etruscan speakers to the North and the inland tribes of Central Italy to the East and South. This political and economic expansion brought them into closer contact with the Greek cities on the Southern coasts of the peninsula and in Sicily. They had long had trading relations with the Greek world directly or indirectly through the Etruscans, but from the third century BCE onward, they became increasingly involved militarily with the politics of the wider Greek world. The growth of their political influence over the Greek cities of Southern Italy resulted first in wars with Illyrians on the Eastern side of the Ionian Sea and with the Carthaginians for the control of the whole island of Sicily, and in the next century in substantial intervention in the political and military conflicts further East. By the first century BCE the Romans were in effective control of the whole Eastern Mediterranean, either directly through the replacement of native rulers by Roman military governors or through a system of clientage that allowed the Ptolemaic rulers of Egypt, for example, to continue under Roman supervision. Cities like Athens,

Antioch or Pergamum exchanged one imperial overlord for another, not always peacefully, as they attempted to maintain as much local autonomy as possible.

But if the political and military advantages lay with the Romans, the small towns of central Italy were inferior in wealth to their Greek counterparts, and they had long followed Greek models as they developed civic organizations. If the linguistic frontier between Greek and Latin speakers remained, it was the non-Greeks who borrowed the more. They had their own healers: the dramatist Aeschylus called Etruria a land of many drugs, and the tombstone of two medical members of the Achonius family at Perugia in the first century BCE is inscribed in both Latin and Etruscan. The famous bronze liver from Piacenza in the Po valley that was used in Etruscan divination implies a certain, albeit limited, knowledge of anatomy, but, as yet, little archaeological evidence has survived to supplement the very scant literary and epigraphic evidence for medicine in the Central Italian countryside. One small Apennine region boasted families of healers, the Marsi, whose reputation as snake-charmers and healers lasted for centuries. They would come down from the hills to Rome, where their expertise in cutting off the heads of snakes to make exotic remedies aroused a mixture of respect and fear. Elsewhere, traveling healers, such as Lucius Clodius of Ancona, will have supplemented domestic remedies, as few towns before the first century BCE will have had a wealthy enough population to have their own civic doctor. The Latin language had certainly developed a sophisticated vocabulary of medical terms in Latin by 230 BCE that allowed an audience to appreciate jokes that derived from earlier Greek comedies, and many Greek-based technical terms were in widespread use a few generations later, when Cato the Elder discussed medical matters in his treatise on agriculture.

This hellenization of medicine is found earliest in connection with religion (details of which were most likely to be written down and summarized later by the historian Livy). The cult of Apollo the healer was introduced officially to Rome in 433 BCE to help deal with a pestilence, anticipating by almost a century and a half the similar arrival there of Asclepius/Aesculapius. After three consecutive years of disease, a priestly consultation of the Sibylline books revealed that only the god could put an end to this epidemic. A formal embassy to Epidaurus the next year received the divine approval, and the god was conveyed to Italy in the form of a snake. He did not come directly to Rome, but landed briefly at the coastal town of Antium, where he took refuge in his temple there and in its sacred grove. After three days, the snake came back on board, only to slip away upon reaching Rome and making his home on the Tiber island, where a temple was erected to him near a sacred spring. The plague stopped 'with miraculous speed'.

The introduction of this alien god was a formal decision sanctioned by the highest civic and religious authorities. It is depicted as a matter entirely between Rome and Epidaurus, and it is only in passing that we hear of a pre-existing cult at Antium in Latium. The Tiber island was also a suitable place for a foreign deity, neither inside nor outside the city. But without the evidence of archaeology we would not know of the similar healing shrine at Fregellae in central Italy, although it may not have been dedicated to Asclepius at this date and its major refurbishment did not take place until the 180s BCE. There were also many healing springs elsewhere in central Italy, where patients left votive models of bodily parts, and a form of incubation may have been practiced at Lavinium, another town not far from Rome. The arrival of Asclepius was thus not just a medical event but also an index of Rome's place within the Greek world, where shrines like that of Apollo Oulios at Elea/Velia attracted many pilgrims to this Italian coastal town and where Southern Italian doctors of eminence, such as Heraclides of Tarentum around 89 BCE, shared in debates with others from as far afield as Alexandria.

The Roman historian Cassius Hemina, writing around 150 BCE, claimed that the first Greek doctor to come to Rome was Archagathus, son of Lysanias, a Peloponnesian, in 219 BCE. This is unlikely, both because the Roman state kept no records of the arrivals of immigrants and because other sources hint at the earlier presence of other Greek practitioners, eager to find patients in an already burgeoning metropolis not far from Greek cities such as Naples. Rather, Hemina's detailed account gives the political and the institutional background to something slightly different: the first civic invitation to a Greek medical practitioner to come and act as a civic doctor. Archagathus was no minor figure, but a distinguished surgeon, some of whose recipes have been preserved for centuries by later authors. He was invited publicly by the Roman senate and given Roman citizenship and a surgery at public expense at a major crossroads, both of which are rewards typically given to a Greek civic doctor. He enjoyed great success at first, but his treatments by cutting and burning gained him the nickname of 'the executioner', and he did not stay long. Pliny the Elder stresses his evil reputation, but Archagathus himself may not have found this frontier city (in Greek eyes) entirely to his liking and left at the expiry of his contract.

Hemina will have been familiar with the public pronouncements of the Elder Cato (c. 234–149 BCE), who made a political career out of a strident hatred of all things foreign. Writing in a letter to his son, around 175 BCE, he warned him against 'filthy Greek doctors', who conspired to murder their patients and to charge them for the privilege. Instead, his son should follow the good, old-fashioned way of healing by which the Roman head of a household, assisted by his wife and bailiff, took responsibility for all those in it, including animals and slaves. How much of his own private notebook of recipes Cato used in his treatise on agriculture is debated, but his remedies there are largely herbal, sometimes accompanied by chants and charms, and even rituals. A cure for stomach worms, based upon pomegranate blossoms, fennel, honey, marjoram and wine, ends with the injunction to the patient to jump down from a square pillar ten times before going for a walk. By contrast with Greek comments about the sacredness of the art, these remedies are accessible and utilizable by anyone, whether herdsman or landowner, for the power to heal resides in the drug and its accompanying rituals, not in the ability of the practitioner. Such ideas lie behind much of what is termed folk medicine and can be found in learned Greek writings as well as in the so-called magical papyri from Roman Egypt.

Cato's dislike of all things foreign was shared by many others in Rome. A group of Greek philosophers were expelled from the city only a few years later: that Greek doctors joined them is unlikely. This corresponded to the greater urbanization of Rome from the second century BCE onward and fears about the depopulation of the countryside and the disappearance of the citizen farmers who had provided the bulk of the Roman army. The unhealthiness of the city was contrasted with the vigor of the countryside (malaria, the scourge of much of Italy until the twentieth century, does not at this date appear to have ravaged Latium or the Campanian plains), the pallor of workers in the mines or in the city slums with the ruddy cheeks of the upland farmer. The contrast between town and country becomes a cliché in Roman literature, and the Catonian appeal to old Roman virtues was reiterated with even greater force and at even greater length by the Elder Pliny, writing his *Natural History* around 75 CE. Prodigiously learned, immensely prolific and for centuries the source for much of Western knowledge of the ancient world, Pliny chose medicine, along with ornamental tables and see-through silk dresses, as the proof of the decline of Roman-life and morals over the centuries under the growing influence of foreigners, most notably Greeks. His history of Roman medicine, a diatribe against Greek medicine and in favor of simple herbal remedies, imposed for almost two millennia the notion of the wholesale inferiority of Greek medicine in Rome to that of their predecessors in the pre-Hellenistic world.

It is an account that, like all good propaganda, contains more than a grain of truth in its saga of sex and skulduggery. The blustery and useless verbiage of the Greeks is contrasted with the practical successes of Roman domestic medicine, the austere morality of the early Romans with the enormous wealth that could be gained by imperial doctors, not to mention their wholesale immorality. Cato's complaints about the Greeks being out to murder their patients were repeated at length, and the low status of their doctors as immigrants, and often slaves and ex-slaves, was emphasized by comparison with the privileges of the true Roman citizen *paterfamilias*. Pliny's apparent authority was enhanced by the support of Galen, a

Greek immigrant himself, who, in an equally rhetorical outburst, denounced his competitors in Rome as incompetent butchers who had left their home towns for the anonymity of the big city where no-one knew them or the medical crimes they committed. A more critical approach to the evidence, as well as an appreciation of its limitations, allows for a more nuanced understanding of one of the most crucial developments in the whole history of medicine, the acceptance of Greek medicine as the dominant strain in Latin medicine and its subsequent development within Western medicine. Without Latin, Greek medicine might have suffered the fate of Babylonian medicine as a mere intellectual curiosity.

The Roman take-over of Greek medicine was assisted, even if not originated, by the presence of wealthy Romans in the Greek world from the third century onward as traders and, still more, as officials of the expanding empire. Some of the supporters of Archagathus already had links with the Greek world, and it became common for leading officials there to hire Greek doctors to attend them and to bring them back with them to Rome. By the middle of the first century BCE, it had become normal for Cicero and his friends to recommend doctors to those traveling to the East, and for Greeks to see attachment to such a magnate as a way of moving to Rome and of access to the enormous wealth that imperial power had brought. At the same time, the mushroom growth of the city allowed many healers to come independently there. By 10 BCE, tombstones record doctors from all over Greece and Asia Minor, and their numbers continued to increase for the next two or three centuries. There was clearly a market for the services they offered, despite the strictures of Cato and Pliny, confirmed, according to the sources, by grants of citizenship by Julius Caesar and, later, Augustus to all doctors working in Rome. The truth of this tradition is hard to confirm from the epigraphic data available, and the decision may have been as much an attempt to control the arrival of immigrants as a long-lasting public health measure.

6.9. Medicine in the Roman world

Pliny's emphasis on the enormous wealth amassed by royal physicians – in Galen's time they were paid a retainer equivalent to the pay of a provincial governor – shows just how significant was access to the imperial court, as well as the ways in which the money was spent. Rebuilding the city walls of Marseilles was a gesture that might have been expected of a local magnate: that it was paid for by a parvenu, however welcome it might be to fellow citizens, was also likely to be a source of resentment to those who regarded themselves as the civic elite. The career of C. Stertinius Xenophon, who served the emperors Claudius and Nero for at least fifteen years, shows the possibilities open to those at the top of the medical tree. A member of a wealthy family from Cos, who claimed descent from two healing gods, Asclepius and Hercules, he became immensely rich. His city practice alone brought him half a million sesterces a year, more than a provincial governor. He accompanied the emperor Claudius on his expedition to Britain in 43 CE, receiving military decorations appropriate for a centurion and superintendent of engineers. He acted as one of the emperor's Greek secretaries and was publicly commemorated as a member of the council of both Claudius and Nero. He retired to Cos – rumor accused him of playing a part in the death of Claudius in 54 CE – where coins and a large number of inscriptions recorded his and his family's benefits to the island. He acted among other things as the head of an embassy that secured the island's freedom from tribute, and, as well as being a priest at the Asclepieion and priest for life of the imperial cult, he was responsible for a major refurbishment of the shrine, including a library, a small temple and an adjacent bath complex with improved plumbing. Other imperial physicians made similarly lavish gifts to their homes and received suitable thanks in return.

But to concentrate on this small elite is to miss one major difference between the Greek world of the Roman empire and its counterpart in Italy and, later on, in the Western provinces. Both parts of the empire acknowledged the Greekness of medicine. The great majority of practitioners known from the West bear Greek names. Some of them or their families came originally from the East, and others may have adopted or, if slaves, may have been given their Greek names by their master, rather as ice-cream parlors flaunt Italian and elegant restaurants French names, regardless of their ancestry. This is a public announcement of the acceptability of their medicine within Roman society, even if it says little about what they believed or practiced or how it interacted with more traditional domestic medicine. But this acceptance came at a price. Imperial physicians aside, only a very tiny number of the names recorded in the West shows signs of wealth. Indeed, until the mid-second century, the great majority indicate that the doctor was not a Roman citizen. Few doctors hold local municipal office or priesthoods, and if they do, they are largely connected with guilds of ex-slaves. The contrast with Asia Minor, where some physicians are connected to the leading families of the provinces, held priesthoods, or donated lavishly to civic organizations, is marked. There are no records of Western doctors traveling, like Galen and several of his Pergamum contemporaries, hundreds of miles to study at Alexandria, or, like Heraclitus of Rhodiapolis in Lycia, gaining a Mediterranean-wide reputation as a poet and scholar as well as a physician. Indeed, where we can follow doctors from one place to another, the trajectory is almost always from East to West, and those who moved the other way, like Galen or Soranus of Ephesus, were returning home with a somewhat jaundiced view of life in the capital. Soranus, around 100 CE, remarked how Roman children were likely to suffer leg deformities through being taught to walk on hard marble floors. This does not mean that a doctor or surgeon in Italy or Gaul could not attain moderate wealth. The house of the surgeon at Rimini and many of those at Pompeii associated with medical instruments are proof enough. But, like the doctors of Roman Egypt, they are far from being among a town's elite and are equally removed from the utmost poverty. At Beneventum in Central Italy, a Greek immigrant made enough money as a doctor to set his family up in some state and for a descendant to hold a high municipal office.

The presence of so many slaves and ex-slaves in the West also marks it out from Greece and Asia Minor. Some of those who came will have been captured in one of the many wars and revolts during the first centuries of Roman rule, others will have been taught medicine in Rome by their master or someone employed to teach them before they could be sent out onto the streets to make money. This must have been profitable for both master and ex-slave, for the Roman lawyers decided that it was illegal for a master, who still retained some rights over his slave even after granting him freedom, to demand that his ex-slave doctor spend all his time treating himself for free – a prohibition that also applied to the master of an ex-slave actor – and thereby deprive him of his livelihood. The epigraphic record is also skewed by the many slaves and ex-slaves who are commemorated in slave cemeteries as doctors, midwives and nurses to the hundreds of slaves in the households of the emperors and their relatives, a feature of Rome in the first centuries of the Empire, although legal regulations for the buying and selling of slave doctors continue to be repeated into the sixth century CE, when their numbers are likely to be small.

This perception of the Greekness of medicine in both East and West will also have been assisted by the growth of Rome itself as a Greek city. By 100 BCE its Greek-speaking population was perhaps second only to that of Alexandria, and many of the upper classes, including emperors like Hadrian and Marcus Aurelius, spoke and wrote Greek fluently. It was a bilingual city, in which Galen could find himself surrounded by friends from home from the moment he arrived there in 162 CE. How much Rome was integrated into the Greek world by this date can be easily demonstrated from two pseudonymous texts in the Galenic Corpus. The Greek author of *Theriac, for Pamphilianus*, studied in Rome with a doctor called Maecius in the 160s. His contemporary, the author of *The Properties of the Centuary*, had moved from somewhere in the Greek world to Rome and reported back to his medical brother Papias on his experiences while attending a Roman doctor on his rounds there. An earlier writer on pharmacology, Scribonius Largus, who also joined in Claudius' British expedition, compiled a surviving collection of medical recipes in Latin, as well as another that, if not written in Greek, was being quoted in Greek by authors from the end of the first century CE onward. The contrast with the centuries of the Republic, when Greek doctors in Southern Italy, like Heraclides of Tarentum, looked to the Aegean world and Alexandria, was marked.

A further proof of the integration of doctors in Rome can be found from the careers of the first doctor in Rome with a Mediterranean-wide reputation, Asclepiades. Asclepiades was an immigrant from Bithynia in North-West Turkey – some believed he came as a captive or a refugee – who had established a great reputation in Rome for his medicine and his ideas in the 90s BCE. Some credited him with nearly-miraculous cures, and his abilities were praised by none other than Cicero. His theories were controversial. His belief that health and illness involved the passage of invisible particles through what he posited as pores in the body was connected by Galen, almost certainly rightly, with the Epicurean philosophers' doctrines of a world made up of atoms and void. (A similar medical grouping came later to be known as Democriteans, thus emphasizing the connection with atomist philosophy, such as was later espoused by the Roman poet Lucretius. Asclepiades' mechanistic and materialistic explanations for the workings of the body, together with his opposition to functional, teleological explanations, also suggest influence from Erasistratus. Nature, he argued, often acted dangerously. He was also famous for his range of therapies, including massage, passive exercise for the seriously ill, and music in the treatment of mental illness, not to mention a liberal recommendation of wine. His therapeutic slogan 'swiftly, safely, pleasantly', a development of ideas found in earlier 'Hippocratic' writings, was as notorious among his later opponents as it was attractive to his patients.

His career marks the point at which Greek medicine was effectively transferred to Rome, for he attracted pupils from around the Mediterranean, including Illyria, Syria and Sicily, and doctors calling themselves Asclepiadeans can be found later in Gaul and in remote Asia Minor. At least one court doctor, Marcus Artorius, in 23 BCE, was an Asclepiadean, as were many of the writers on pharmacology whom Dioscorides drew on for his *Materia medica* eighty or ninety years later. In some Greek handbooks he is regularly cited as the last representative of the so-called Dogmatic or Rationalist physicians before Galen, for he derided the Empiricists' belief in observation alone as the key to therapy. In his view, once one had decided by careful observation whether an illness was the result of an imbalance, a blockage or a flood in the atoms and pores of the body, it was easy to devise a therapy that would eliminate the cause without the need for complicated nosology or symptomatology.

His ideas were extended by his pupil Themison (active around 60 BCE) and still further expanded by another immigrant from the East, Themison of Tralles (active around 60 CE), into what became known as the Method, and its adherents as Methodists. The history and dating of early Methodism are obscure, but Themison seems to have modified Asclepiades' views in two ways. The cause of any illness could lie in either the atoms or the pores. If the channels were too large or the corpuscles too small, the result was 'looseness'; if the reverse, 'stricture'. There was also a third intermediate state, neither one nor the other, but different from what was healthy. Diseases could be either acute or chronic, the former being the result of 'stricture', the latter of 'looseness', which Themison explained further in what was said to be the first treatise with this explicit organization. He may also have divided treatment into three universal stages: an initial increase, a middle period where things remained constant, and a final stage where the disease diminished. He may also have been among the first, if not *the* first, to have stressed the importance of the *diatritos*, a recurrent three-day-period, as marking possibly significant stages in treatment.

Even more famous was Thessalus, a charismatic practitioner, who attracted crowds of followers and who addressed letters in approximately 60 CE to the Emperor Nero declaring his superiority to all his competitors. He modified the Method still further so that, he claimed, it could be learned in a mere six months, proof, thought Galen, of the lack of education and sense of social responsibility on the part of the slaves, ex-slaves and down-and-outs who were attracted to Methodist medicine. His major innovation seems to have been to enlarge and then subdivide the 'common features' of all illnesses. He distinguished dietetic from surgical common features, dividing the latter into external, e.g. a thorn, and internal, the latter further subdivided into those involving an abnormality of place (e.g. a fracture), of size (e.g. a tumor), and of insufficiency (e.g. a fistula or ulcer), each of which required different treatments. Careful observation allowed for the modification of treatment as the cure progressed, even switching to an apparently contrary therapy if the situation warranted it.

Galen's attacks on the Methodists, these 'Thessalian asses', and his criticisms of their apparent illogicalities should not be allowed to mislead. They were not interested in Galenic logical subtleties, and at times their actual therapies differed only slightly from those of both Galen and the Empiricists. Even Galen was prepared to acknowledge, at times, a later Methodist, Soranus of Ephesus, for his changes to the Method, although he says little about Soranus' most famous work, his handbook of gynecology. They achieved considerable success in Rome, where Galen was forced to share his duties as an imperial physician with Methodists like Statilius Attalus, and the surviving Latin medicine of late-Roman North Africa is heavily influenced by Methodist doctrine. Its main advantage was that it was a medicine for the big city. It provided clear guidelines for practitioners, emphasizing claiming an easy link between observation and therapy and concentrating on generalities rather than on spending a long time on the individual encounter between doctor and patient. Celsus more than once drew attention to the distinction between a doctor practising in a *valetudinarium*, a slave hospital, or a large city and one who attempted to understand and prescribe for the individual. Better, he thought, to pay more attention to fewer patients than to treat all and sundry with a therapy that might not be targeted precisely. But, on the other hand, an ordinary patient might well prefer a swift diagnosis and treatment that might work well over a lengthy and expensive attempt to bring one back to one's natural balance of humors. Even Celsus was prepared to admit that if one did not know one's patient, proceeding by means of common features was sensible, but it was far better to have a friend as a doctor than a stranger.

6.10. Doctors in the Roman Empire

The picture of medicine in the first and second centuries CE in Rome and the Greek East is one of considerable variety, as practitioners jostle with each other for patients with a variety of therapies. Some, particularly in a large city, specialized in certain activities, such as oculists or specialists in fistulae; others, the so-called 'crowd-pullers', attracted patients through spectacular performances in the marketplace. Some gave public lectures and demonstrated their surgical skill by deploying flashy instruments, while others wrote up their ideas for transmission to friends and potential patients, in the hope of having them placed, like some of the recipes of Marcellus of Side, in a public library by imperial fiat. Tiberius Claudius Menecrates, around 50 CE, based his new creation of a 'clear and logical sect' on the 156 books that he had written. Others, such as Heraclitus of Rhodiapolis, 'the Homer of medical poetry', Thrasippus of Corinth, 'second to none in Greece for his medicine and his poetry', and Hermogenes, the doctor-historian of imperial Smyrna and the author of 77 books of medicine, made considerable play with their cultural and intellectual superiority. Some found new ways of announcing their abilities. At Ephesus, during the two days of the Great Festival of Asclepius, there were medical contests among elite physicians whose results were inscribed in stone at the local Hall of the Muses. Others marched in public at festivals in honor of Asclepius or Apollo as a way of displaying their professional acceptance. It was a world filled with medical practitioners, and potentially so expensive for cities employing civic doctors that a series of emperors from the late first century onward attempted to reduce the numbers entitled to tax privileges to those whose expertise and moral standing was approved by the city council and within limits on numbers depending on the town's size. In larger cities, so varied was the competition between healers that, if a patient had not already been recommended a doctor by family or friends, the choice was far from easy. Galen, typically, wrote a treatise advising the potential patient what should be done, setting out the questions that might be asked to sort out the ignorant from the competent, as well as the conditions he expected a doctor to be able to treat successfully. He also implied that there were others, himself included, who could do more than this (to him) basic minimum.

But this picture applies largely to Rome and the Greek world. Elsewhere, similar evidence is largely lacking. Little direct information, for example, survives for Roman Alexandria, except for Galen. By contrast, from chance finds of papyri we know more about the life of doctors in small Egyptian towns and the ways in which traditional Egyptian medicine was gradually assimilated within an overall Greek framework. Very few physicians were among the wealthiest members of the community, but they were not among the poorest, with some owning land in addition to what they could make from their practice. Most owned at least a few books, many written by distinguished earlier physicians, but others containing recipes or instructions for surgery from otherwise unknown authors. Just as medicinal substances were transhipped from ports on the Red Sea across the desert to the Nile and then down to Alexandria, so in turn books could travel from the Mediterranean world via Alexandria to towns higher up the Nile. A book by Galen could be read there relatively soon after it had been composed in Rome.

If Egyptian papyri give a glimpse into the daily life of a doctor as reader, prescriber and tax-payer, our information on the practice of medicine in the Western and Northern provinces of the Roman Empire comes entirely from archaeological finds either from excavations or from inscriptions, supplemented by scattered literary references. They reveal a preponderance of healers with Greek names, some expressing themselves in Greek, but almost nothing directly about their theories. The first medical treatise that can definitely be associated with that region, a drug handbook by Marcellus of Bordeaux, was written only in the late fourth century. It incorporates several local remedies, much as a similar contemporary book from Crocodilopolis in Egypt includes both native and Greek recipes. Other writings from North Africa in this period and later also show much Methodist medicine, sometimes mixed with Hippocratic or Galenic ideas. Finds of surgical instruments, ranging in date from the Stanway hoard from Britain just before the Claudian conquest to the fifth century, show that practitioners could obtain, or have specially made, some elegant and sophisticated instruments, although few, perhaps, would have had a collection as large as that of a doctor at Rimini in 257–8 CE that included over 150 instruments, many only slightly different from others that he would have used for similar operations. Some towns acted as regional medical centers. There was an association of teachers and doctors at Aventicum in modern Switzerland, a conjunction often found in Asia Minor, and Bordeaux in the time of Marcellus had a reputation as an intellectual center. The distribution of so-called oculist's stamps, used to brand sticks of ointment used to treat eye diseases, suggests that Bar-le-Duc in France, for example, served as a base from which healers could go out to visit the countryside. The presence of surgical kits alongside some of these stamps indicates that these practitioners treated more than just eye conditions and that their expertise need not have been restricted to surgery or ophthalmology.

6.11. Military medicine

Most information on Western provincial medicine, however, comes from military sites in frontier regions. Until the late first century BCE, most campaigns were relatively short and close to home. Athens employed doctors for the fleet and for individual campaigns, and the less severely wounded could be brought home. Alexander the Great solved the problem of dealing with the sick and those no longer able to serve on his campaigns in the Middle East by founding settlements where his Greek veterans could live in retirement and also protect the distant troops in the case of revolts. A variety of historians and military theorists emphasized that it was a mark of a good general to look after his troops, even having his own physician treat some of them in his own marching tent. But the campaigns of Emperor Augustus in the last decades of the first century BCE required a different and more structured response. The Roman army was now a permanent force, operating hundreds of miles from home with a mixture of citizen soldiers and non-citizen auxiliaries. They could no longer rely, at least at first, on friendly locals to house any wounded, and garrison towns, often called 'coloniae', were not created immediately. The solution was for the army to develop its own medical corps, no longer dependent on doctors on short-term contracts, who worked primarily in the great legionary fortresses and in a few forts. Each legion had its quota of medical staff, including bandagers and some trainees, and there was a similar organization in Rome for the Praetorian Guard and the police and fire-brigade. The fleet also had its own medical staff, probably each doctor serving on one ship, and there was a similar, although less generous, provision of medical assistance for the auxiliary, non-citizen units.

Each legionary fortress had its own hospital, *valetudinarium,* possibly on the model of earlier slave hospitals. By 50 CE it took a recognizable form in the West (much less is known about the Eastern frontier fortresses): a large rectangle with a central space and small rooms on either side of a corridor running around the whole building. There was often a large hall by the entrance, where operations might be performed, and in some hospitals a small shrine to the healing gods, often containing votive dedications by the doctors. Marching camps would have temporary buildings, as in the first stage of the fortress at Inchtuthil in Scotland, and a few auxiliary forts might also have a space given over to a hospital. But it would be wrong to believe that those imposing structures were of much use in war. They were often situated many miles away from the scene of any fighting, and any seriously wounded would have died long before reaching them. They were there to provide assistance to the legionaries in peace time, for roster roles show many of them suffering from a variety of ailments, particularly diseases that spread through contact, like conjunctivitis, and only rarely from injuries suffered in training. Doctors may well also have attended the sick women and children in the settlements that grew up outside the fortress walls, but their numbers were too small and their locations too remote to have been major agents in the transmission of learned medicine to the provinces. Nor need they have remained long there after the ending of their period of service. The Rimini doctor, who seems to have seen military service, returned to a comfortable life in a large house in Italy.

How long these great military hospitals continued in existence depended on circumstances and military necessity. Inchtuthil, a legionary fort in Scotland, was deliberately abandoned only a few years after it was begun, and before a proper water supply could be created, when the Romans decided to withdraw from all of Central and Northern Scotland. Elsewhere, the change from the later third century CE onward to more mobile central forces, with the frontier in part garrisoned by soldier-farmers, saw less and less value in permanent frontier fortresses, and they gradually fell into disuse, as the forces themselves reverted to the older habit of leaving the wounded in friendly towns.

Archaeology has also revealed the importance of religious healing within the various regions of the Roman Empire. All the major shrines of Asclepius at Cos, Pergamum, Epidaurus, Aegae and Lebena underwent massive refurbishment, which often included a theater and a library to accompany the incubation chambers. In some areas, Asclepius could join forces or amalgamate with an original local healing god; in others, it was local deities who attracted worshippers even to remote shrines like that at Carrawburgh on Hadrian's Wall in North Britain or that of the three matron goddesses at the sources of the Seine in Central France. Other deities spread more widely. The snake god Glycon, who came to prominence first at Abonouteichos on the Southern shore of the Black Sea, was widely invoked against the Antonine plague in the 160s, perhaps even as far away as London. Many homes had their own small healing shrines, and patients regularly used amulets and religious spells to ward off or to cure disease, although few went as far as the orator Aelius Aristides in the mid-second century CE, who spent years at the shrine of Pergamum and who described his relations with Asclepius in a long series of orations.

6.12. The career of Galen

One famous doctor who believed that his whole career had been determined by the intervention of Asclepius was Galen of Pergamum (129–216/7), the most prolific and most influential of all ancient medical writers. Born into a rich family with landed estates, he enjoyed an excellent education in philosophy and the Classics before Asclepius advised his father that he should become a doctor. At the age of 17, when many young men were already practicing, he embarked on a medical odyssey that took him to study at Pergamum, Smyrna (Izmir), and for several years at Alexandria. He was brought up in the Hippocratic tradition of humoral medicine by teachers who had themselves studied at Alexandria and had participated in a revival of animal dissection that, according to Galen, had begun 'in the time of our grandfathers'. After a period back in Pergamum, where he gained the position of doctor to the gladiators (a post that brought him into touch with the provincial elite) through an unscrupulous display of surgical expertise, he moved to Rome in 162 CE. He made his name by public lecturing and by some flamboyant displays of anatomy, including some remarkable investigations into the brain and nerves. These performances, initially in public, were later carried out in private before an audience that included senators and even imperial relatives. But they also incurred the hostility of those whose shortcomings were trumpeted in public and assailed by Galen's remorseless logic and unscrupulous rhetoric. In 166 CE, mindful of the murder of a similar young medical genius, or so he says, he slipped away from Rome, returning home to do his 'usual things'.

He was not forgotten. In 168 CE, he was summoned by the emperors to join them in camp in Northern Italy in preparation for a campaign against German invaders. Traveling by sea and land, through frost and snow, he reached Aquileia, only for Emperor Verus to die suddenly in early 169 CE, probably of a heart attack rather than 'plague', and for his co-emperor Marcus Aurelius to return hurriedly to Rome. Galen came later, along with some of the troops. When asked again to go on campaign, he invoked the intervention of Asclepius, and the generous Marcus Aurelius gave him the task of overseeing the health of the obnoxious young Commodus in Rome and in nearby palaces. From now on, perhaps until his death and with a possible brief return to Pergamum, he remained in imperial service under a succession of emperors. This was not particularly onerous, as he seems to have spent much of his time writing (or dictating) many books on anything from lexicography to logic, as well as innumerable volumes on medicine. He also seems to have had a busy practice in Rome, treating slaves and senators, professors and athletes injured in the gyms of Ostia, as well as advising others by letter (one of which, to the father of an epileptic teenager, still survives).

The accession of Commodus in 180 CE began a dark period in his life. Friends disappeared or were executed, and courtiers, himself included, lived in daily fear of their lives in what he regarded as the worst reign in history. He may have wished to withdraw to his villa on the Bay of Naples, but before he could arrange for copies of the books in his library to be sent there, disaster struck. A great fire in the center of Rome in 192 CE destroyed the stores where he kept his books, instruments and valuables of all kinds, as well as the royal libraries from which he might have obtained copies. He had to rely on friends for what they could provide (in some instances only several years later) or rewrite himself. Only a few weeks later, Commodus was dead, and after a civil war, Severus became emperor, once again employing Galen as a court doctor. Whether outsiders saw Galen as the innocent bystander in politics he claims to have been, given the reputation of doctors, before and since, for political intrigue, is doubtful. We cannot know when he composed a treatise on *Slander*, in which he talked of his own life, although its very existence implies that not everyone shared his high opinion of himself. When and where he died is unclear, although later chronographers and Arabic biographers who place his death at the age of 87 are more plausible than a Byzantine lexicon entry that has him dying at 70. Later legends have him buried in Sicily near Palermo or at Perama in Egypt on a pilgrimage to the Holy Land; a final return to Pergamum is plausible but unproven.

It is only too easy to take Galen at his own valuation, to approve of his insistence that the practitioner must exercise thought as well as skill and to appreciate his achievements as an anatomist and acute medical observer. He was a success in his own lifetime, even if he was not as unique a court physician as he would wish us to believe, and we have more contemporary records of him and his influence (for good or ill) on other medics, philosophers and even Christians than for any other practitioner, save those he attacked. Recent research has shown that he was not alone in many of his activities, whether as anatomist, philosopher or public intellectual – he himself would have rejected the appellation of 'sophist', although his rhetorical skills and his willingness to pronounce instantly on almost any topic would surely have qualified him for that title. His ability, even today, to provoke his readers to praise or, more often, anger gives us insight into how he would have appeared to contemporaries. But it should also serve as a warning that one should not consider him as a typical ancient physician. He stands at the apex of a pyramid of ancient healers, distinguished from the start by his family wealth and his education, and surrounded by friends and contacts at the very top of ancient society. The gulf as he acknowledged occasionally, between him and the poorer doctors to whom he gave recipes, books and instruments was wide, and that between a doctor in Rome, Pergamum and Ephesus and one in rural Gaul, upland Italy or the Nile valley is likely to have been wider still. To begin a study of ancient medicine by taking him as in any way typical is a failure of method.

6.13. Medicine in the Late Antiquity

The medical world of Rome and the Greek East, as described by Galen and evinced by inscriptions and other archaeological findings, is distinguished by a vibrancy and a variety that is rarely found elsewhere. The absence of any formal legislation or qualifications allowed almost anyone to offer healing, although women were more often derided for the lack of literacy (with some rare examples) and for their reliance on magical or disgusting remedies, like Scribonius Largus' old woman from Crete with a hyena-skin remedy for epilepsy. Medicine was an art practiced in public, not only in lecture halls and open markets but even in the sick room, which was often crowded with friends and visitors eager to offer their own suggestions or simply to see the doctor at work. To succeed required public skills: the art of persuasion and appropriate behavior, to avoid frightening, annoying or even shaming the patient by revealing their buttocks or private parts. In major cities, one's cultural qualifications might come into play, to supplement or replace any knowledge gained from long acquaintance with the practitioner and his family.

150 years later, the picture had changed considerably and continued to grow darker. Our information from the Latin provinces declines in both quantity and quality, as barbarian invasions, a worsening climate and the gradual decline of civic institutions led to progressive impoverishment. By 400 CE, all of the intellectual institutions of the Greek world that had sustained a long medical education such as that enjoyed by Galen and his friends had disappeared, except in North Africa. When a choice had to be made of what deserved to be copied and recopied, short practical handbooks and useful remedies took precedence over theoretical discussions or learned commentary. Traffic in medicinal substances endured, and tombstones recording doctors were still to be found occasionally, even in far-distant Wales, but we know little about the intellectual context in which they were to be found.

After a gap of almost a century in our surviving medical writings, we enter into a very different world, which can loosely be called Galenism. The great encyclopedias by Oribasius in the fourth century and Aetius of Amida in the sixth, as well as the major compendium of Paul of Aegina from the seventh, show a progressive adherence to Galen. The percentage of entries ascribed to Galen grows steadily, as sections from other authors in the Hippocratic tradition, including his teachers, are either left out or subsumed under his name. The vigorous debates between the sects of the first and second centuries are replaced by summaries, compendia and commentary (something also evident in law and literature). Not all of this can be put down entirely to the powerful rhetoric of Galen and the feeling that, after the harvest he had reaped, there was little but stubble to be harvested. Animal dissection, let alone human dissection, disappears from the records, and brief summaries of Galen's conclusions (some of them his own) replace his detailed instructions about how to dissect. Other authors, such as Magnus (around 350 CE), who was given a publicly funded auditorium at Alexandria, brought together ideas on themes that Galen had discussed at various points across his writings, such as uroscopy and sphygmology, into elegant treatises, thereby giving them a clearer profile, and perhaps an added significance, than existed previously. Lecture commentaries on Hippocrates on the Alexandrian model were supplemented or replaced by lectures on Galen, all presuming an acquaintance with Aristotelian philosophy and logic.

It would be foolish to speculate what brought about this change, but its consequences are clear. First, it imposed a Galenic template on intellectual medicine, not just in ensuring that it was his view of the body in sickness and in health that prevailed but also in making crucial choices as to which books should be copied and preserved. Many of Galen's own books were never recovered after the fire of 192, and others seem to have been quickly lost. Those of others, supporters of Hippocratic medicine as well as Galen's opponents, were thrust aside unless they filled in a gap left by the master. This explains the survival of Soranus' book on gynecology, that of Aretaeus (perhaps late 2nd century CE) on nosology, Rufus of Ephesus (around 100 CE) on jaundice and on how to question a patient, and Dioscorides on medical botany, themes touched on only briefly, if at all, by Galen himself. Rufus of Ephesus, who was highly regarded by Galen, suffered particularly badly: only fragments survive of his huge compendium of medicine *For the Layman* or, its alternative title, *For Those Who Have No Doctor to Hand* (a situation perhaps even more frequent in Late Antiquity, to judge from the commonplaces on this topic in some late prefaces). Surgery seems to have continued to develop, especially at Alexandria, for the surgeon who traveled West from Alexandria to treat Innocentius, a friend of St. Augustine in Carthage in approximately 390 CE, was said to have had a great reputation simply from studying there. Paul of Aegina was famous as an expert on obstetrics, and his compendium of medicine included the finest account of ancient surgery. It went beyond Celsus or the surgery described in the pseudo-Galenic *Introduction* in the number of operations it mentioned, and showed how an Alexandrian surgeon might take into account Galen's anatomical discoveries in performing a tracheotomy. But it also meant that Galen's view of Hippocrates and Hippocratic medicine came to be normative, despite the variety of opinions known to Galen in his Hippocratic commentaries. Even today, the history of ancient medicine tends to be written through Galenic spectacles, and only recently has attention been drawn to the value of the pseudo-Galenic works by Methodists and others that survive only because they were thought to be genuinely his.

Increasingly, to judge from what has survived, commentary replaced original writing in a society where acquaintance with certain books, whether in law, religion or medicine, came to be the mark of the true (or simply the elite) expert. Galen (and probably his teachers also) had singled out certain writings associated with Hippocrates as significant and more worthy of comment than others. How far his prejudices were followed in the classroom is unclear, but a similar, well-structured syllabus of Galenic books had certainly been formed by 500 CE in Alexandria, if not much earlier. It began with some basic introductions, an overall view of medicine and its terminology, followed by simple student introductions to anatomy and a brief guide to diagnosis from the pulse and from urine. From these a student would move on to read Galen's account of the make-up and workings of the body and learn about disease before finally studying his big books on the method of healing and, although this is less certain, preventative medicine. This 'syllabus' was among the first medical documents translated into Syriac, the vernacular language of the Middle East, and subsequently into Arabic and, ultimately, Latin in medieval Europe. A few lectures on some of the books of Galen and Hippocrates were also given in Latin at Ravenna, an outpost of Byzantium in Northern Italy, and perhaps also in Rome, where there was still a substantial Greek community, but no further. The opening texts in this syllabus were the most studied and were soon provided with student summaries and study aids, thus making it possible to examine and distinguish between practitioners on the basis of their understanding of certain texts. (A similar concentration on the correct interpretation of canonical texts can also be observed at the law school in Beirut, not to mention in the religious debates that periodically engulfed Christianity.)

This continuity is particularly evident in the legal regulations for doctors, often repeating in the sixth century and beyond administrative rulings for the titles and remuneration of doctors that had been issued four or more centuries earlier. Rules for prices for medically trained slaves were reiterated long after all other evidence for slave doctors had disappeared. But law is a conservative profession, and large cities like Constantinople, Rome and Alexandria will have retained many of their institutions long after those in smaller cities had disappeared and the records of winners in the medical contests at Ephesus had become incorporated into the walls of a great Christian church.

It is Christianity that above all marks off Late Antiquity from the age of Galen. The Gospels and the book of Acts, its foundation documents, are full of references to the healing miracles of Christ and his Apostles. They not only demonstrate the power of the Christian God over all others, but also show how these miracle stories could be used as a means of converting unbelievers. Galen himself had shown an unusual interest in Jews and Christians, approving of their high morality but regarding as illogical their belief in miracles, for why should a Creator suddenly at a whim change the world he had so carefully organized? Some clerics, and especially bishops like Gregory Thaumaturgus ('the wonder-worker'), Bishop of Neocaesaraea in Northern Asia Minor from 238 to around 270 CE, gained a reputation as healers, and crowds flocked to the tombs of these healing saints to find relics that they might take back home with them to protect against disease, rather as they had made pilgrimages to shrines of Asclepius. Sometimes the opposition between Christians and non-Christians involved the destruction of a well-loved shrine. The great temple of Asclepius at Aegeae in Southern Turkey was razed to the ground in 331 CE, its columns broken up or reused in a church on the site, an insult to the thousands who had once come there to be healed. Not surprisingly, when, in 363 CE, the Emperor Julian wished to restore the older cults, this was one temple that he particularly wished to rebuild. At Menuthis near Alexandria the shrine of Isis was transformed during this period into a church, but this did not stop pilgrims from coming to receive healing and to consult the oracle of Isis. In response, the Bishop of Alexandria fostered (or invented) the cult of two Christian healers, Saints Cyrus and John, whose 'surgery' was deliberately placed across the road from the site of the shrine. Not until 483 was the decision taken to act against the Isis worshippers there, and all trace of the temple was removed after a riot.

The Christianization of healing was a long process. Around 450 CE, Bishop Theodoret in central Syria was convinced that Asclepius was still being worshipped in the villages in his large diocese. The philosopher Proclus could still visit the shrine of Asclepius at Athens as late as 470 CE to pray for the recovery of a young girl, but it was demolished only a few years later. Half a century later, Bishop Caesarius of Arles was deeply worried about a countryside where folk medicine and superstition were rampant, while the expensive medicine of the town he considered tinged with paganism. He may not have been wrong for the 'religion of Asclepius' continued to have many adherents among learned doctors, especially at Alexandria. The Neoplatonist philosopher Damascius, writing his *Life of Isidore* at the end of the fifth century, included many doctors among his heroes. Professor Gesius at Alexandria, 'whose rhetorical abilities removed all difficulties of exposition', openly doubted the miracles of Saints Cyrus and John, pointing out their origin in Hippocrates or Galen. His punishment was to be afflicted with a disease that no mortal doctor could cure. Only after a full and contrite confession of his mistake, so

Christians boasted, did he recover through the divine powers of the Saints he had ridiculed.

But for many of the sick, the major change brought about by the advent of Christianity was the provision of care for those in need, one of Christ's injunctions in the Gospels to his followers. It was a concern that extended beyond the initially small number of Christians to embrace the entire community, as Emperor Julian acknowledged in a letter (whose authenticity has been challenged), in which he encouraged the civic authorities under his renascent pagan rule to imitate the Christians in using charity as a means of conversion. In the Greek and Roman ideology of euergetism, the wealthy provided benefits of various kinds for the members of their communities in a form of patronage, but although they might pay for food during a famine, healthcare did not figure largely, if at all, in their benefactions. It was perhaps the Jews who first created institutions to provide accommodations and assistance, including medical assistance, for all their co-religionists, particularly the many pilgrims to the temple at Jerusalem. Similar buildings, probably from the imperial period, can be found near the shrines of Asclepius at Cos and Epidaurus, and some cities paid for the service of a doctor during their major festivals. But institutions and buildings that dealt mainly with the sick first appeared in two different contexts. By the late first century BCE the supply of slaves obtained by conquest or trade was beginning to decline, and thus, it has been suggested, it became advantageous for slave owners to look after their sick slaves somewhat more carefully in special barrack blocks. Celsus, as we have seen, thought this to be at best a stop-gap medicine, and it may not always have involved a doctor rather than the wife of the bailiff. At about the same time, permanent hospitals were being erected in legionary fortresses, reflecting the imperial concern for the well-being of the troops as insurance against mutiny. One might also think of the shrines of healing gods as being somewhere that would provide succor for the sick, proto-hospitals. How medical this might be was revealed by recent excavations at Rhodiapolis and Aezani, both in Turkey and dating from the second century CE. Medical instruments were found at Rhodiapolis in a building adjacent to a temple of Asclepius erected by the wealthy doctor Heraclitus, but this may simply have been an extension of the common custom of doctors to treat some of their patients in their own home. At Aezani, excavators found a large building with very many small rooms that was next door to, and may even have formed part of, the refurbished shrine of Asclepius, which was well known to Galen for its healing spring. That surgical activity was carried out in some of the rooms is clear from the large number of instruments found there, and one can imagine that some patients came to the town to take advantage of both religious and secular healing.

But it was the Christians who most extended charity to the wider community, first by providing assistance in plagues and famines to the sick in their own homes (and occasionally in the house of a bishop) and then, after the legalization of Christianity in the fourth century, in specific buildings. The variety of names used to describe them, such as sick house, poor house, hostel for strangers, old-folks home and hostel, indicates a range of overlapping activities, depending on the needs of their locality. They were erected first in the Greek world. Around 350 CE Bishop Leontius set up a series of hostels in Antioch and at the neighboring spa of Daphne. A few years later, another bishop in Turkey, Eustathius, built a 'poorhouse', where those crippled with illness could find succor. His more famous contemporary St Basil of Caesarea erected 'almost a new city' outside the walls of the city where the sick, the poor, the leprous and the stranger could be cared for and receive medical assistance. The defensive tone in which his work is described suggests a certain opposition to his plan, and a friend tells how he was moved by the sight of large numbers of lepers, half-dead with ulcers and sores, trying to force their way into churches and temples. By the end of the century such 'sick houses' were common enough to form the basis of an extended metaphor in a bishop's sermon. Under Eastern influence they spread to Italy, first in Rome in 397 CE, followed by Ostia, and then to cities in North Africa. Fewer are known from Northern Italy, and fewer still from North of the Alps, possibly a reflection of the economic and social dislocation there in the fifth and sixth centuries rather than an absence of evidence.

By contrast, these institutions were ubiquitous in the East, even in remote places, and often no more than a room off the courtyard of a church or monastery. Local church regulations enjoined the provision of such hostels and laid down rules to ensure their financial stability. Around 500 CE, the city of Edessa in Southern Turkey had two or three such institutions, all of them small, in a community of 8,000–10,000 souls. In emergencies such as famine or plague, they were supplemented by beds set up in colonnades to house the hundreds of sufferers who had flocked thither from the countryside. As time went on, some hospitals became larger and more complex. Ephesus in 420 CE had one with 70 beds, Jerusalem in 550 CE one with 200 beds, and St. Sampson in Constantinople may have been larger still. Signs of simple specialization also appear. By 500 CE, Edessa had a hospital for women, and by 600 CE big hospitals at Antioch and Constantinople were divided into male and female wards. By 640 CE surgery was being carried out at St. Sampson's, and a special section was given over to patients with eye diseases. Physicians were in attendance at a variety of hospitals – at Oxyrhynchus in Egypt one medical family even ran its own small hospital – and some provided treatment as well as nursing care for the sick.

But to seek a purely medical hospital is a work of supererogation. Smaller hostels may have provided no more than a bed and simple food, while those in Rome, Constantinople and the Holy Land may have specialized in looking after pilgrims. Neither names, statutes nor legislation, which are far more concerned with financial than medical problems, can easily reveal the extent of medical involvement or the opposition to it. The Statutes for the famous theological school at Nisibis in 476 CE ordered that sick students were to be cared for by their fellows in their own cells. 75 years later, a hospital was founded at the school with money from the King of Persia and provided with every necessity so as to shift the burden of care from students to an official. It was also to stop the students from being 'plundered and dishonored' when they went into town in search of more sophisticated assistance. Theological students were forbidden to associate with physicians, 'for the crafts of the world are unworthy to be read with the books of holiness', and those who did were only allowed back if they were of good character and natives of Nisibis. In all likelihood, then, this was a hospital without doctors: others elsewhere, no less Christian, will have followed a different line.

6.14. Conclusion

To trace the history of medicine in Antiquity over more than a millennium and a half is no simple task. There are huge gaps in our information, as well as a substantial geographical range. Even concentrating on Roman military medicine involves bringing together documents and archaeological information from as far afield as Scotland (where epigraphy reveals supplies of wine and blankets) and the borders of Wales (where a learned doctor inscribed a Greek poem in honor of Asclepius and Panacea) to the deserts of Africa, where pupil bandagers are recorded on the roll of the legion at Lambaesis. The medicine of the small town in the countryside is almost invisible, whether in Central Italy or in Asia Minor, although life there is regularly contrasted with that of the big city such as Rome, where latrines, workshops and market stalls sent their filthy waste into the Tiber to poison the fish that had swum freely in the upper reaches of the river. Studies of skeletons at Pompeii and Herculaneum have shown that their inhabitants enjoyed a relatively healthy diet, with plenty of fish and, according to Galen, excellent wine in the vicinity. But one may hesitate to generalize about the health of the population as a whole from these small and relatively prosperous maritime communities.

It is tempting in such circumstances to try and bring together this scattered evidence into a smooth narrative, or to concentrate on a few places (Athens, Rome and Alexandria) or types of medicine, such as that of the army or civic doctors. But even here there are huge gaps: Hellenistic and Roman Athens are far less studied than the golden years of the fifth and fourth centuries BCE, and military medicine after 250 CE is almost as obscure as it was three hundred years previously. Religious medicine has also tended to focus on the cult of Asclepius, to the neglect of other more local cults and of Christianity, whose texts are more abundant and more varied than those of their non-Christian predecessors. This is also to underplay what is the most striking feature of ancient medicine: its abundant variety. This was aided by an absence of effective legislation relating to medicine. The Roman legislation on tax immunities neatly shows some of the difficulties. The repetition of roughly the same legislation over several centuries indicates the clash between civic authorities, who are always eager to economize, and those who offered healing but whose abilities could not be neatly defined in the absence of any agreed criteria. Some specialists, such as dentists or specialists in the treatment of fistulae, could be considered in the same way as physicians, but not those who healed by chants and exorcisms even if there were those who swore that they had benefitted thereby. Even military experts might reckon that regular training and drills were far more conducive to good health than the services of doctors. Not until the very end of Antiquity with the development, perhaps first at Alexandria, of something resembling a syllabus, can we find a widespread acceptance of the understanding of specific books as a criterion of excellence, for Galen's insistence that his ideal doctor should be acquainted with the tenets of a select few 'ancients' is typically partial.

Our appreciation of the development of medical ideas is similarly flawed, for Galen's insistence on the superiority of 'his' Hippocrates led to the eventual disappearance of almost all alternatives to his preferred authors and ideas. Only the chance discovery of a papyrus from Roman Egypt more than a century ago allowed us to glimpse the existence in fifth- and fourth-century Greece of vigorous debates about the causes of illness, encouraging the move away from a concentration on certain texts that Galen had associated with his hero, with a consequent lack of interest in most of the texts within the so-called Hippocratic Corpus. Investigations into pseudo-Galenic writings, as well as fragments preserved on papyri and, with Rufus of Ephesus, the survival of some writings in Arabic translation, are beginning to remove Galen from his position as the self-imposed arbiter of all that was significant in ancient medicine.

To warn of the weakness of the evidence for Greek and Roman medicine is not to deny the attempt to bring to light the medicine of a distant era. The medicine of many other societies and other periods is even more restricted in its evidential bases and in its historical context. Ancient medicine also follows the survival pattern of almost all other aspects of Antiquity, with its written evidence heavily concentrated in three periods: the fifth and fourth centuries BCE, the first century BCE to the third century CE and the Christian centuries from the fourth century to the sixth century. This temporal coincidence does allow historians to discuss medicine in its wide social and intellectual context, and in a way that has still to be done for most medicine before the Renaissance. The variety that is becoming still more evident as more is discovered about Classical Antiquity, whether in its medical ideas or in its relationships between practitioners and their societies, offers new insights into the past as well as points to ponder in the modern world.

6.15. References

Bliquez, L.J. (2015). *The Tools of Asclepius. Surgical Instruments in Greek and Roman Times*. Brill, Leiden.

Boudon-Millot, V. (2012). *Galien de Pergame : un médecin grec à Rome*. Les Belles Lettres, Paris.

Bouras-Vallianatos, P. and Zipser, B. (2019). *Brill's Companion to the Reception of Galen*. Brill, Leiden.

Craik, E.M. (2015). *The Hippocratic Corpus. Content and Context*. Routledge, London.

Harper, K. (2021). *Disease and the Course of Human History*. Princeton University Press, Princeton.

Harris, W.V. (2016). *Popular Medicine in Graeco-Roman Antiquity: Explorations*. Brill, Leiden.

Jones, A. and Taub, L. (2018). *The Cambridge History of Science I. Ancient Science*. Cambridge University Press, Cambridge.

Jouanna, J. (2017). *Hippocrate*. Les Belles Lettres, Paris.

Jouanna, J. (2024). *Médecine et tragédie en Grèce antique. Scripta minora 1961-2023*. Les Belles Lettres, Paris.

Nutton, V. (2016). *La Médecine antique*. Les Belles Lettres, Paris.

Nutton, V. (2020). *Galen: A Thinking Doctor in Imperial Rome*. Routledge, London.

Pormann, P. (2018). *The Cambridge Companion to Hippocrates*. Cambridge University Press, Cambridge.

Singer, P.N. and Rosen, R. (2024). *The Oxford Handbook of Galen*. Oxford University Press, Oxford.

Von Staden, H. (1989). *Herophilus: The Art of Medicine in Early Alexandria*. Cambridge University Press, Cambridge.

7

Greek Medicine: Semiotics, Physiology, Nosology

Amneris ROSELLI
Università degli Studi di Napoli L'Orientale, Naples, Italy

7.1. Texts and tradition

The reconstruction we are able to make of Greek knowledge of the human body and physiology significantly depends on the nature of the texts available and the alterations in their transmission. Today, we have a large number of complete, anonymous medical texts dating from the middle of the 5th to the 4th centuries BCE, forming part of the Hippocratic Collection[1], but we have almost entirely lost outputs from subsequent physicians until the 2nd century of the Roman Imperial Age, when the impressive corpus of Galen's writings appeared, a reservoir for earlier literature and the repository of his own doctrines, which were heavily dependent on earlier sources[2]. For philosophical literature, although many writers have dealt with animals and, above all, human nature and physiology before Plato, Aristotle and Theophrastus, we can read many fragments, but we do not have complete texts[3].

1 For general information on the treatises in the Collection and on Hippocratic medicine in general, see Jouanna (1999), Craik (2015) and Pormann (2018).

2 General information on the doctrines and writings of Galen (129–216) can be found in (Hankinson 2008a); a rich profile of the history of ancient medicine built entirely on the sources can be found in Nutton (2024). See also above, Chapters 5 and 6.

3 Reference edition, see Diels and Kranz (1951–1952).

History of Life Sciences and Medicine 1,
coordinated by Lorenzo PERILLI.

Aristotle's biological and psychological works are the focal point in this field[4], but much information is also derived from later doxographic sources[5]. The output of the philosophical schools of the Hellenistic age is known only through fragments reconstructed from sources of the Imperial Age (Plutarch, Galen, doxographies, commentaries). The presentation of opinions on anatomophysiology, humoral theory, nosology, semiotics and etiology of diseases therefore requires us to follow a road made up of several parallel and sometimes convergent paths, in some sections very well lit but in other sections rather dark, as well as to rely on the testimony of elliptical and not always objective sources in the presentation of their predecessors' ideas. Here, we attempt to present the elements of continuity and convergence of thought on these issues and, by means of a few texts, to illustrate the most general ideas at the heart of the dialog between ancient authors over the centuries.

7.2. The living body

In the Heroic age, the bards, who in the *Iliad* sang of the terrible days of battle and the immense slaughter of the Greek and Trojan heroes, referred to the viscera and fluids that come out of gashes of wounds with a crudeness that can be likened to some of today's war films; blood splatters, brains, marrow, bones and excrement are described in their appearance and colors with a kind of fascinated disgust. Violent death, caused by the crushing and smashing of skulls, the exposure and spilling of viscera and the amputation of limbs, accompanied by the exhalation of the last breath, offer a privileged moment for insight into the mechanisms of life in Homeric epics and have influenced the formation of body images destined to endure.

Later, through the dissection of animals such as the goat, the keener eye of the researcher was able to observe the channels that connect the brain to the outside world and that were most likely the channels of sensation and respiration; this is the case for Alcmaeon of Croton, who lived around the beginning of the 5th century BCE[6], and likely others such as him, on whom we have no information.

Empedocles was one of the first writers to give us a rapid sketch of the body's internal structure, in the few verses in which he described the mechanism of

4 *Timaeus* plays an important role in Plato's doctrines; to Aristotle, we owe the books on the *History of Animals*, the *Parts of Animals*, the *Generation of Animals*, other minor works on animals, *On the Soul* and *Parva Naturalia*; to Theophrastus, the books *Historia Plantarum*, *On the Causes of Plants* and *On Sensation*.

5 On doxography, see Mansfeld and Runia (1996).

6 See Alcmaeon, No. 24 Diels/Kranz.

respiration[7]. Following the widespread analogical method of explaining non-visible phenomena with known visible phenomena, Empedocles compares the movement of blood and air in the veins to the alternating movement of air and water in a clepsydra. Empedocles, who writes in verse and uses sustained language, calls the blood *phonon* ("the product of kill") and *teren haima* ("delicate blood", probably alive, fragile in its freshness), the air *aither* and the veins that carry it *liphaimoi surigges*, "channels devoid of (or deprived of or with little) blood", or simply *diodoi* ("passages"):

> It is in this way that all inhale and exhale: for all channels of flesh, which the blood leaves (*liphaimoi surigges*) extend to the surface of their bodies. And at the openings (*epi stomiois*), the furthest limits of their skin (*rhinon*) are perforated through and through with dense furrows, so that the blood (*phonon*) lies hidden, while easy access is cut by these passages (*diodoisi*) for the air (*aitheri*). When from here the delicate blood (*teren haima*) then rushes backward, the air (*aither*), boiling, rushes after it in a raging surge, but when it (the blood) leaps back, the other is exhaled again – just as when a child plays with a clepsydra of handy copper [...] (Translation by Laks and Most).

A few other fragments tell us that Empedocles defines the liver as "rich in blood" (B 150) and thought as "blood placed around the heart" (B 105): air and blood thus appear as two constituent components of the living body, and the heart and liver as the two deputies with a special relationship to blood.

When we move away from poetic language, these channels of air and blood are generally referred to as veins (*phlebes*); this is what Diogenes of Apollonia (second half of the 5th century BCE) called them a few decades later[8]. According to the account by Simplicius (6th century CE), "thought derives from the fact that air together with blood occupies the whole body through the *phlebes*", of which – Simplicius adds – Diogenes provides "an accurate anatomical description"[9].

The description of the vascular system is recurrent in the medical and philosophical literature of the 5th to 4th centuries BCE; it can be found in different redactions in various treatises by Hippocrates: *On the Sacred Disease* (Chapter 3–4 = 6–7 Jones), *On the Nature of Man* (Chapter 11 = *Nature of Bones* 9), *Epidemics* II 4, 1 (= *Nature of Bones* 10), and again, as witnessed by Aristotle's *Historia Animalium* 511b-513a, in the writings of other ancient physicians and

7 See Empedocles B 100 Diels/Kranz.

8 Diogenes of Apollonia fr. 64 Diels/Kranz.

9 Simplicius, *Commentary on Aristotle's Physics* CAG 9, 153, 14–16 Diels.

philosophers: Syennesis of Cyprus, the aforementioned Diogenes of Apollonia, and Polybus, who, according to Aristotle, is the author of Chapter 11 of *On the Nature of Man*; after reviewing his predecessors, Aristotle in turn offers his own description of the vein system. This is obviously a "topical" piece in the literature on the constitution of the human body, expressed through a rather fantastic anatomy. In most cases, the veins start at the head and run down through the trunk toward the extremities of the limbs, with paths described partly on the basis of observation of extremely thin animal and human cadavers and partly speculatively; however, in *On the Sacred Disease*, veins run up from the liver and spleen to the head, and similarly, in *Epidemics* II, 4 1, where the "liver vein" (*hepatitis*) is described as being raised from the liver; Aristotle's description concentrates on the heart. Why is there so much emphasis on the path of the veins? This is because the vessels ensure the passage of fluids and air through the body. The brain, heart, lungs, liver and kidneys function as nodes, collecting, processing and distributing air and fluids, primarily blood. The body is permeable and maintains continuous exchanges with the outside: food, drink and air inhaled mainly through the nose and mouth are necessary for the preservation of life and the accomplishment of the organism's functions, as they nourish it and regulate its temperature. Inside the body, food is transformed: the residues of its transformation are eliminated through excretion and various other forms of emission of very subtle waste products, such as exhalation, perspiration and sweat. Interruption of this continuous process of exchange between the outside and inside causes illness or death.

Aristotle's biological writings (notably *Historia Animalium, De Partibus Animalium*) represent the most advanced stage of knowledge in the field of pre-Alexandrine anatomophysiology; Aristotle describes the structure of the trachea and esophagus, the different forms of the stomach in different animals; the textures of the brain and medulla, liver and kidneys, lungs and heart; the membranes lining the different cavities; and the parts destined for reproduction and those for waste evacuation. Aristotle's biological writings cover the entire living world (human and animal) and represent a powerful synthesis, extraordinary in its breadth and detail, destined to survive beyond even the most advanced Alexandrine research, which, as far as we know, is strongly focused on human anatomy and innovative in its research method: it was at this time that the transition from anatomical knowledge, by accumulation of mostly occasional observations, to a systematic study of the body took place.

The few and varied details of anatomophysiology in the 5th–4th centuries are enough to show that the conformation of the internal parts of the human body, and the properties of the fluids that move within them, are largely the product of a vivid imagination and explain why a common model could not be found; however, they show that, over the course of time, physicians and "philosophers" have constantly attempted to explain physiological phenomena in terms of "fluid dynamics", sometimes using comparisons with plants or animals (e.g. in embryological treatises such as *On Generation, On the Nature of the Child*), sometimes using mechanical models such as the clepsydra (Empedocles, B 100) or communicating vessels (see section 7.4).

The richest documentation, the one that allowed us to learn about the human body and its functions, provided by medical literature, consists of fully preserved treatises from the middle of the 5th century onward; however, these treatises transmit earlier material and knowledge that had already been organized and rewritten over and over with a certain degree of freedom[10]. Medical texts speak of the body from a different point of view than that of epic poets or naturalist philosophers: for the physician, the body is not the one seized at the moment of death, nor even the product of investigations into nature, but the much more usual and longer observable body of the patient, which is unable to perform some of its functions (from the most insignificant alterations in sleep to difficulty in breathing, to the inability to speak, digest, defecate or reproduce), a body that suffers in different ways and in different parts, producing particular manifestations such as fevers, tremors, shivers, sweats, spasms, bleeding, vomiting, extreme weakness, agnosia and delusions. Every sick body offers a wealth of observational data over time, and the elaboration of this innumerable quantity of data, which are all grouped under the category of symptoms (the word used is "signs", *semeia*), has required generations of physicians to develop selection criteria (clarifying what makes a certain phenomenon a *semeion*) and probabilistic models (Di Benedetto 1966) that would then enable them to formulate diagnoses and prognostic indications. The hypothesis, which is not always explicit in medical texts, that enables diagnosis and prognosis is the cause-and-effect relationship that links each symptom to one or more of the solid, liquid or gaseous components that are present in the body. However, the sources relating to philosophers tend to emphasize the contribution that distinguishes one philosopher from another (this also depends on the fact that we can only read fragments and doxographies, and not full texts); medical literature, in its earliest phase, reveals itself to be the product of a great collective enterprise, which only sees the emergence, from the last decades of the 5th century onward, of individual personalities who express themselves in the first person while remaining anonymous.

10 In recent years, research has highlighted the similarities between Eastern medicine, particularly Babylonian, and Greek medicine by comparing the pathology manuals produced in the two traditions; one of the first books of modern conception on this subject is Horstmanshoff and Stol (2004) (see also Geller 2010, and above, Chapter 2).

7.3. Nosology and symptoms

The earliest medical texts in the Hippocratic Collection (*Diseases* I, II, III and IV, *Affections*, *Internal Affections*) consist of lists of juxtaposed nosographic notices, with a minimal number of first-person interventions by their authors, or rather editors or compilers. These treatises are anonymous and, when studied comparatively, show various successive levels of reshaping. They show that by the middle of the 5th century BCE, a format had been imposed for the writing of reports devoted to the various diseases, which, in this form, also had to be memorized; this format was modified and updated over time – our documentation allows us to observe the changes that took place in less than a century, between the middle of the 5th century and the first decades of the 4th century – without ever losing its essential characteristics[11]. The greatest novelty is the inclusion in the body of the individual nosographic records of a generally not very extensive section devoted to etiology, which is absent from the earliest texts; this testifies to the emergence of an interest in the "explanation" of symptoms and allows us to glimpse, albeit still confusedly, the profile of the different doctrinal trends and authors; the etiological section is generally placed after the list of symptoms. A typical nosographic record thus contains symptoms, etiology, therapy and prognosis, the latter in variable order, sometimes mixed together. The most constant element, at the beginning of each record, is the list of the particular and distinctive symptoms of each disease. The name of the disease, if present, occupies the first position, but it is not an indispensable element. Indeed, in most cases, the disease dealt with in the record can be identified only by the position it occupies in a sequence that regularly lists diseases in order, from head to toe (an order already perfectly documented in Mesopotamian medical manuals); paradoxically, a record taken out of context may not provide sufficient indications for disease identification.

Symptoms are often very numerous, and in the nosological treatises they occupy, on average, the space of a few lines. The fact that the signs recognized as characteristic of a certain disease do not always appear all together, but some seem to predominate while others do not, produce a multiplication of homonymous diseases, with each "variant" considered a disease in its own right. Already in the last decades of the 5th century, the multiplication of diseases was seen as an element of theoretical weakness by previous physicians, to whom the author of *On Regimen in Acute Diseases* opposed the need for a selection of relevant symptoms, which would induce a less compartmentalized reading of nosology (see *On Regimen in Acute Diseases* 3):

---

11 So much so, that innovations in the format have made it possible to establish a relative chronology of nosological treatises (Jouanna 2009).

> But though they wished clearly to set forth the number of each kind of illness their account was incorrect. For the number will be almost incalculable if a patient's disease be diagnosed as different whenever there is a difference in the symptoms, while a mere variety of name is supposed to constitute a variety of the illness (Translation by Jones).

The physical and psychic manifestations that are interpreted as symptoms result from the observation of numerous cases. Sometimes they are the result of observations that have been collected over a fairly long period of time, as the following two records show:

> *Diseases* II 50: If the pipe (*surinx*) of the lung becomes aphthous, there is violent fever, pain in the middle of the chest, and itching of the body; the patient's voice is hoarse, and he expectorates thin moist sputum, *sometimes*, though, the sputum is thick and like barley-water. A heavy smell arises in the patient's mouth, like that of the raw fish, and *from time to time* there appear in this sputum hard pieces like the fleshy excrescences of an ulcer. (The patient) becomes lean in his upper parts, and his whole body is very wasted. His cheeks blush, and *after a while* his nails become curved, dry and yellow-green (Translation by Potter).

> *Diseases* II 68: Livid disease: a dry fever comes on, occasional shivering, the patient suffers pain in his head and inwards parts, and he vomits bile; when the pain is present, he cannot look up, but feels weighed down. His belly is costive, and his complexion, lips, and the whites of his eyes become livid; he stares as if he were being strangled. *Sometimes* his color changes too, and turns from livid to yellow-green (Translation by Potter).

In the first case, the disease is identified by the ulcers that occur in the upper part of the trachea and is indicated by a conditional proposition ("if the pipe... becomes aphthous"), the sores being located in the canal leading to the lung[12]. In the second, the name of the disease is expressed by an adjective, "livid", its location is missing, but from its position in the list, it is clear that it is considered a disease of the abdomen. Although there are several chapters introduced by a simple noun, such as *phrenitis*, *tuphos* (which is not our typhus) and *pthisis* (which is not modern consumption), the list of diseases identified by a noun is still quite short compared with that of diseases introduced by a conditional proposition, which is also an indication of the unsteadiness of the technical vocabulary (Roselli 2018, 2019).

---

12 The same term *surinx* is used by Empedocles B 100.

The first symptoms in both cases are fever and pain, which are not always localized, and, in the first and second cases, itching all over the body and chills, respectively. These are followed by symptoms of excretion (sputum, spit, vomiting), impaired function ("cannot see", "has pain in viscera") and an altered appearance (eyes, cheeks, lips, general complexion, nails), which are assessed by smell (bad breath) and color (especially in the case of the latter disease, which owes its name to its livid color but also in the case of the former). In both cases, the symptoms must have been observed by different people over different periods of time. Qualification is important (fever is "high/low", "dry"; sputum is "thin", "moist", "thick"); sometimes, as in the second record, comparisons are made with phenomena of everyday experience to better define their particularities (density of sputum: "like barley water"; appearance of sputum: "with lumps like the fleshy growths of an ulcer"; "odor of sputum": "odor like raw fish"). In some cases, the symptoms reflect the patient's own statements ("he cannot see, but he feels heavy").

The identification of symptoms reaches the highest level of detail in the clinical histories of the seven books of *Epidemics*, where the presentation is less "discursive" and tends to become a list of nouns, verbs and adjectives juxtaposed without any syntactic link; above all, in *Epidemics*, the symptoms concern single clinical cases, which implies a more attentive evaluation of individual manifestations. Data are not recorded or selected insofar as they can be linked to a pathology, but insofar as they help to define it; they are arranged in an orderly chronological sequence, even including details of the times of day and night. The text of each "clinical case" consists almost exclusively of symptoms, but summary data also appear, as in the case of a young woman who had a difficult birth, transcribed at the beginning and end of the record in *Epidemics* III 1, 12 (77–79 Jouanna):

> A woman who lay sick by the Liars' Market, after giving birth in a first and painful delivery to a male child, was seized with fever. From the very first there was thirst, nausea, slight pain at the stomach, dry tongue, bowels disordered with thin and scanty discharges, no sleep.
>
> *Second day*. Slight rigor; acute fever; slight, cold sweating around the head.
>
> *Third day*. In pain; crude, thin, copious discharges from the bowels.
>
> *Fourth day*. Rigor; general exacerbation; sleepless.
>
> *Fifth day*. In pain.

Sixth day. The same symptoms; copious, fluid discharges from the bowel.

Seventh day. Rigor; acute fever; thirst; much tossing; towards evening cold sweat all over; chill; extremities cold, and would not be warmed. At night she again had a rigor; the extremities would not be warmed; no sleep; slight delirium, bur quickly was rational again.

Eighth day. About mid-day recovered her heath; thirst; coma; nausea; vomited bilious, scanty, yellowish matters. An uncomfortable night; no sleep; passed a copious discharge of urine, I did not see it.

Ninth day. General abatement of the symptoms; coma. Towards evening slight rigor; vomited scanty, bilious matters.

Tenth day. Rigor; exacerbation of the fever; no sleep whatsoever. In the early morning a copious discharge of urine with sediment; extremities were warmed.

Eleventh day. Vomited bilious matters, of the color of verdigris. A rigor shortly afterwards, and the extremities became cold again; in the evening sweat, rigor and copious vomiting; a painful night.

Twelfth day. Vomited copious, black, fetid matters; much hiccoughing; painful thirst.

Thirteenth day. Vomited black, fetid, copious matters; rigor. About mid-day lost her speech.

Fourteenth day. Epistaxis, death.

The bowels of this patient were throughout loose, and there were shivering fits. Age about seventeen (Translation by Jones).

This record is very similar to that of a modern medical record; many symptoms are recurrent; the doctor who records the data, even with the help of the woman's carers throughout the day, nevertheless implies a selection of the phenomena observed and the existence of a grid according to which he questioned the patient or her entourage. The fact that the doctor writes that he did not see the urine on the eighth day justifies the absence of any information on its appearance, as is the case for the urine on the 10th day ("with sedimentation"); in short, he is filling in an empty box.

In the clinical cases of *Epidemics*, many symptoms can be expressed not only by name (e.g. thirst, nausea, shivering and insomnia) but also by adjectives designating the patient as "thirsty, comatose, nauseous, shivering, insomniac, aphonic" (*dipsodes*, *asodes*, *komatodes*, *phrikodes*, *agrypnos*, *aphonos*), whereas in nosological treatises, adjectives, as we have already noted, refer only to symptoms.

7.4. Humors

In the nosological treatises, the second part of each record is devoted, from a certain point onward, to etiology. This is where humors are often mentioned, although they are also sometimes referred to in the symptoms section to define the appearance of excretions and dejecta through adjectives ("phlegmatic, bilious" and others). The mention of humors in the etiological section does not explicitly state the "system" of humors accepted by the writer of the record, assuming that there is one; it can at least be said that, in the earliest nosological treatises, the two humors, bile and phlegm, are considered the most directly involved in explaining the various pathologies, with blood also playing an important part.

The system of the four humors ("blood", "phlegm", "yellow bile" and "black bile"), which, owing to Galen's important patronage, became ubiquitous in the Hippocratic-Galenic medical tradition right up to the modern era, is in fact only one of the "systems" we know and certainly does not represent the perspective that was commonly accepted in the 5th century. Attested for the first time in the Hippocratic treatise *On the Nature of Man* (late 5th–early 4th century), and probably "created" by its author (Polybus), it is the highly speculative product of fixation in a quaternary pattern – which would prove extremely effective over time – of doctrines rich in variants, which attributed a role in maintaining bodily functions and producing disease to an indefinite number of different fluids. Acknowledging that there are four constituent humors permanently present in the body, albeit in different quantities, makes it possible to continue the doctrine of the four Empedoclean elements ("air", "water", "earth" and "fire"), specializing in the microcosm of the human body, by replacing the primordial elements with humors that can be observed in some way. The four humors, which are related to the four temperaments ("warm", "cold", "dry" and "moist"), are subject to alterations in quantity and quality, and can be associated with the four seasons – an important fact for a medicine that takes account of the external environment – and with the four ages of man (in this case, it is difficult to establish whether it was rather the existence of a fourfold system that led to the ages being segmented in this way). Health and disease result from the balance and imbalance that occur in the mixture of humors under the influence of food, drink, exercise and the air we breathe, which modify their quality and quantity (*On the Nature of Man* 4).

The extraordinary success of this scheme has obscured not only other doctrines that can be reconstructed only today owing to their use in the treatises that presuppose them, but also the account given at the opening of *Disease* IV of a different quaternary system that potentially had the same chances of success. In addition to blood, phlegm and bile, the *Diseases* IV system provides a watery humor (Greek *hydor* or *hydrops*) that prevents the splitting of bile into "yellow" and "black" bile. The two treatises *On the Nature of Man* and *Diseases* IV are two roughly contemporary documents (second half of the 5th century BCE) that approach the problem of body constitution from the same perspective and explicitly base it – as well as, simultaneously, health and disease – on the notion of balance between humors with opposing properties and efficiencies (warm, cold, dry and moist). *On the Nature of Man* underlines the link between humors, seasons and ages and thus the link between health and disease, with seasonal and environmental factors and the nature of individuals in their life cycle; *Diseases* IV highlights the role of humors in the generation of individuals, linking their health and disease to the humoral constitution of their parents. *Diseases* IV also explains the link between humors and parts of the body using the notion of "source", the place from which each humor springs or gathers in greater quantity, implying that each is found in some way throughout the body: blood has its source in the heart, phlegm in the head, bile in the liver and *hydrops* in the spleen, an organ that plays an important role here, particularly in pathology (and which Aristotle considers as a mere counterweight to the liver, failing to recognize its other, more important functions). This quaternary system also identifies four routes for the evacuation of excess humors (nose-mouth, bladder, intestines, skin), which has a practical effect on therapy, as it identifies the places to which accumulations of humors, or their evacuation, must be moved. In these early chapters of the treatise (whose chapters are numbered continuously from *On Generation* and *On the Nature of the Child*), we find a comprehensive, although brief, account of the physics of fluids inside the body, illustrated by the communicating vessels analogy: when one cavity empties, other fluids flow into it or are drawn to it. According to Lonie (1996, p. 300), "a humoral theory results in a speculative anatomy"; this view of body cavities and sources also helps to explain preferences for certain foods and drinks on the basis of the need for the four "sources" to correct imbalances between humors (see *Diseases* IV, Chapter 39):

> Each of these sources, since they are small and in the body's interior, indicates in virtue of its own peculiar power to the rest of the body

> what in each case is bilious or phlegmatic or blood-like or watery in what is eaten or drunk[13](Translation by Lonie).

The system based on the four sources of the four humors, according to the author, enables the body to implement a continuous process of self-regulation that ensures the preservation of health. It is worth reading at least a few extracts from chapters 32–41 of this little-known text, to which all physicians of the Hippocratic lineage could subscribe, and which offers a summary of physiology and nosology while explaining when and how food and beverages are metabolized:

> *Diseases* IV: 32. Both woman and man contain in their bodies four varieties of humor: it is from these that diseases originate, except for those diseases which are the result of violence. These varieties are *phlegm, blood, bile and water*: they contribute to the sperm neither its smallest nor its weakest part, and when the animal is born it contains as many varieties of humor, both healthy and unhealthy, as its parents had. […]. 33. First, then, I intend to show how an excess or deficiency of bile, blood, water or phlegm occurs. It originates from food and drink, in the following way; the stomach, when it is filled, is the source of all the body's nourishment; but when it is empty, it draws upon the whole of the body, which thus wastes away. There are also four sources, from which each of these humors goes to the body, when the sources have drawn from the stomach; and these sources in turn, when they are emptied, draw upon the body. Also the body draws, independently, upon any contents there may be in the stomach. Now, the source of blood is the *heart*; of phlegm, the *head*; of water the *spleen*; of bile, *the receptacle upon the liver* (i.e. the gall bladder). These are the four sources of these humors, apart from the stomach. Of them, the most hollow are the head and the spleen, this having the most space in them – but I shall show this more fully later on. […]. 34. The body draws through the sources I have mentioned from the food and drink entering the stomach, each humor attracting its like through the veins. […]. 40. The head and spleen on the other hand are particularly prone to disease: they are affected both by their *natural humors*, when they receive an excessive amount of these, and by the other humors as well. For they have numerous thick channels leading into them, and these are of a very venous and hollow kind, so that they have space for the other humors as well, as it enters them little by little

13 This implies a much more refined sensitivity than that of the body itself, crudely understood as the body with which we feel hunger and thirst.

> and mingles with their *natural humor*. Just as there is more space in a large vessel than in a small one, so it is with the head and the spleen: they of all these reservoirs have the most space. When their veins become filled with humor, then from this humor disease arises in them. [...] 41. I have shown that there are four humors which cause harm to man, and that they have four sources. Now there are four ways by which a man purges himself of each: the *mouth*, the *nostrils*, the *rectum* and the *urethra*. When there is an excess of one of these humors, and the humor is causing distress, then if a man is purged by one of these passages, no illness from the humor oppresses him (Translation by Lonie).

Since the humors are subject to quantitative and qualitative variations (not only of the four basic temperaments – warm, cold, dry and moist – but also of many others, such as sourness, acidity, bitterness, sweetness, salinity, etc.) and tend to overload one another, their balance is precarious, and health, the result of their equilibrium in the mix, is an ideal state that is always sought after but never attained in a stable manner. The body is constantly deviating from this ideal point, and medicine becomes the necessary and principal tool for restoring and maintaining it through the use of various dietary, pharmacological and surgical instruments at its disposal.

7.5. Pneuma

In addition to humors, a ubiquitous and, in some treatises, highly valued constituent of the body is *pneuma*, a term that not only designates wind and breath but also inner air, respirations and, in the medieval Latin tradition, "spirits". For these notions, we also use the more generic terms air (*aer*) and breath (*physa*), which even lends itself to a treatise (*On Winds*) but later specializes in the pathological sense of "gas, flatulence". Notably, as early as the Homeric poems, the breath that leaves the body (then called *psyché*), when one takes one's last breath, was even understood for many centuries as the sign of death. Ancient medicine's highly successful whole-body formula *ekpnoon kai eispnoon* ("breathing out and breathing in"), which appears in *Epidemics* VI 6, 1, sums up the idea of breathing involving the whole body and extending through the pores of the skin (Debru 1996). It has already been mentioned that pneuma moves through the same channels as blood, and it is not difficult to imagine that, like blood and other humors, it plays a role in physiology and pathology.

The following passage, taken from *On Regimen in Acute Diseases (Appendix)* 7, explicitly illustrates a case of altered air (and fluid) mechanics and its consequences:

> Stoppage of air (*pneumaton*) through the vessels (*phlebes*): when pains come on, affluxes (*epirrhusies*) of *dark/black bile* (*melaines choles*) and *sharp fluids* (*drimeon rheumaton*) are occurring; the patient suffers gnawing pains in his inward parts. The vessels (*phlebes*), being *irritated* and too *dry*, stretch tight, swell up, and attract the fluxes (*ta epirreonta*); from this, since *the blood becomes disordered* and the *air* is no longer able to follow its normal paths (*hodous*) through the blood, chills occur as a result of stasis, along with darkening of the vision, loss of speech, heaviness of the head, and even convulsions, if (the fluxes) have already reached the *heart, liver, or vessel* (the aorta?); from this, patients become epileptic or are paralysed, if the fluxes (*ta rheumata*) invade the surrounding parts and are *dried up* because air cannot pass through[14] (Translation by Potter).

This passage concerns the diagnosis of influxes of acrid and bilious humors, which cause air to stop moving through the vessels. The sign of an influx of black bile and acrid humors (which are not to be understood here as constituent humors of the body but as pathologically produced humors) is pain, perceived by the patient as something biting inside the body; the localization of the pain in the veins is not made by the patient but by the writer of the text, who hypothesizes that the influx of these humors stretches the veins as in a spasm, inflaming them so that they no longer allow the pneuma to move. Stoppage causes cooling. A series of symptoms, such as blurred vision, aphonia, heaviness of the head and spasms, occur when the sour and bilious flows reach the heart and liver and, probably, the aorta (however, the word used here, *phleps*, does not allow us to clarify this point); the author has clearly adopted a doctrine of body function centered on the blood and the heart, which has been identified as the center of the sensory and intellectual faculties. Further propagation of corrupted flows in the areas surrounding the liver and heart produces epileptic-type phenomena. In such cases, therapy consists of evacuating the sour and bilious flows mixed with the blood by means of bloodletting and purgative drugs.

These types of texts, in which theoretical premises are assumed but not spelled out, are accompanied by more explicit texts that offer an explanation of the functioning of the body and disease that is equally accessible to non-experts: this is the case with *On the Sacred Disease*, a treatise that, in many ways, belongs to the same environment in which the doctrines of Diogenes of Apollonia were asserted, focused on air as the primary element of all beings (see section 7.2). *On the Sacred Disease* also attributes a key role in the movement of pneuma in the body, and notes

---

14 See also the explanation of epileptic seizures in Chapter 14 of the treatise *On Winds*, which links the disease to the ingestion of food and drink that alters the properties of the blood, as the production of pneuma by the blood causes the vessels to clog.

the pathological effects of its blockage[15]. To provide an explanation that can be understood by a non-specialist audience, while enhancing the role of physicians over that of a variety of imposters, the author of *On the Sacred Disease* provides his description of the path of the vessels in the body and explains the role they play in determining disease, in that they are the pathways for air.

> *On the Sacred Disease* 4 Jouanna = 7 Jones: By these veins we take in the greater part of our breath, for they are the vents of our body, drawing the air to themselves, and they spread it over the body in general through the minor veins and cool it; then they breathe it out again. For the breath cannot rest, but moves up and down. If it is caught anywhere and rests, that part of the body where it rests becomes paralysed. A proof is that should minor veins be so compressed, when a man is lying or seated, that the breath cannot pass through the vein, a *numbness* immediately seizes him (Translation by Jones).

On the Sacred Disease attributes epileptic seizures to the cessation of pneuma movement caused by phlegm, while *On Regimen in Acute Diseases* attributes them to the cessation of blood movement caused by pneuma. For one, the main organ is the head, for the other the heart. Both texts share the idea that the free circulation of humors through the body (in particular, blood, which has a privileged status in relation to the others) and pneuma is the condition of health, whereas the factors involved are at least comparable, the role played by air and blood in the different treatises is divergent and will remain so for many centuries. Physiological explanations are not the result of accumulated knowledge, and even improvements in anatomical knowledge were not enough in antiquity to definitively impose one explanation over the other.

7.6. Permanence and innovation in Hellenistic and imperial medicine

The Hippocratic texts make no clear distinction between arteries and veins, which are anatomically different but are designated by the same name (*phlebes*), seemingly because they all carry blood, other fluids and air (only Praxagoras of Cos [4th–3rd centuries] distinguished between the two)[16]. In Hellenistic and early

15 This is a very old belief in medical thought, as shown by some fragments of Praxagoras of Cos (see below 7, 6, n. 16).

16 Lewis (2017) offers an up-to-date study of pneuma, both by translating and commenting on Praxagoras' fragments (pp. 1–212) and by discussing his doctrines on arteries, pulse and pneuma (pp. 213–310).

imperial period, Erasistratus[17] and the Erasistrateans distinguished between *phlebes*, which carry blood and possibly other humors, and *arteriai*, which carry air from the outside, to the lung and heart, before being distributed throughout the body. The observation of cadavers seemed to prove this distinction of functions, since in the dissection of a dead animal, the arteries are devoid of blood. Praxagoras' arteries narrow toward the extremities of the body until they become *neura* (nerves) and are, in fact, the instruments of motion that drive the heart. Similarly, the term *neuron*, which in the earliest texts refers generically to cord-like structures, tendons and ligaments connected to the muscular system, takes on a more precise meaning as a result of anatomical research; a system of nerves can now be perceived, originating in the head, to which a task must be assigned (in accordance with the Aristotelian principle that "nature does nothing in vain"); for knowledge of nerve anatomy, the sources recognize Herophilus, the other great Alexandrian anatomist (3rd century BCE), as the first[18]. According to Erasistratus, the interweaving of the three different vessels (*triplokia*, interweaving of arteries, veins and *neura*), which cannot be observed because they are too thin but can be assumed on the basis of reasoning, characterizes the structure of all tissues (what Aristotle had aptly called "homeomerous", homogeneous parts), and is roughly equivalent to the tissues of modern terminology. In this way, Erasistratus is able to explain the presence of blood and pneuma, now specialized as vital pneuma and psychic pneuma, in all parts of the body, albeit in differing quantities.

Attempts to explain the physiology of living organisms in mechanical terms, considering the body as made up of atoms in motion but lacking qualities of its own, have been made on several occasions since Hellenistic times; in the field of medicine, the most important names are those of Erasistratus (3rd century), as we have seen, the great anatomical investigator in Alexandria, and Asclepiades of Bithynia (1st century BCE), whose achievements in the field of therapy in Rome have been favored in the sources. Asclepiades was the originator of methodical medicine, which, as far as we know, was hugely successful in the early imperial period. The explanation of physiology and pathology in terms of rarefaction and condensation, obstruction and the filling of voids postulated by representatives of the methodical school – notions that have always been present in medical writings but did not exhaust the range of causes among which qualities dominated – as well as their renunciation of anatomical knowledge, imposed a radical disruption with the

17 Erasistratus' fragments are published by Garofalo (1988).

18 See the edition and commentary on the fragments by von Staden (1989), with a rich review of Alexandrian anatomy.

past; this would have required breaking medicine away from its deepest roots and the imagery it had built up from body observation, which in fact did not happen.

The physics of solids and fluids in the living beings had developed in response to the demand for explanations of the various manifestations of disease and continued to emphasize the notions of mixing, balancing and the equilibrium of qualities, attributing them to the different components of the body. At the same time, the study of food (meat, cereals, vegetables, etc.) had given rise, over time, to widespread classifications of different foods according to their properties of heating and cooling, drying and moistening, stimulating what is sour or sweet, etc., classifications that would not have been easy to abandon. Galen had no difficulties placing the Hippocratic doctrine of *On the Nature of Man* at the heart of his "physics", as being the closest to the physician's experience and the most suitable to his needs (the two works explicitly devoted to this theme are *On the Elements According to Hippocrates* and *Commentary on the Hippocratic Treatise On the Nature of Man*) (Hankinson 2008b). The best evidence for this hypothesis seems to be provided by the short but highly successful *On the Anomalous Dyskrasia*, in which Galen classified the "bad temperaments" at the root of disease on the basis of the four temperaments of humors ("warm", "cold", "dry", "moist") and their combinations ("warm and dry", "warm and moist", etc.), while making use of the advanced knowledge of the anatomical structure of tissues and parts acquired in his day. In this treatise, Galen applies the notion of "bad temperament" to the study of inflammations and fevers, two pathologies that made up a very large proportion of known diseases at the time, demonstrating that pathological observation and medical tradition remain central to the interpretation and "construction" of the physics of the living body (Garcia Novo 2010).

7.7. References

Boudon-Millot, V. (2012). *Galien de Pergame. Un médecin grec à Rome*. Les Belles Lettres, Paris.

Craik, E. (2015). *The Hippocratic Corpus: Content and Context*. Routledge, London.

Debru, A. (1996). *Le corps respirant. La pensée physiologique chez Galien*. Brill, Leiden/New York/Cologne.

Di Benedetto, V. (1966). Tendenza e probabilità nell'antica medicina greca. *Critica storica*, 5, 315–368.

Diels, H. and Kranz, W. (1951–1952). *Die Fragmente der Vorsokratiker*, 6th edition. Weidmann, Berlin.

Garcia Novo, E. (2010). *Galen. On the Anomalous Dyskrasia (De inaequali intemperie)*. UCM/Editorial Complutense, Madrid.

Garofalo, I. (1988). *Erasistrati fragmenta*. Giardini, Pisa.

Geller, M. (2010). *Ancient Babylonian Medicine: Theory and Practice*. Wiley Blackwell, Chichester.

Hankinson, R.J. (2008a). *The Cambridge Companion to Galen*. Cambridge University Press, Cambridge.

Hankinson, R.J. (2008b). Philosophy of nature. In *The Cambridge Companion to Galen*, Hankinson, R.J. (ed.). Cambridge University Press, Cambridge, 210–241.

Horstmannshoff, H.F.J. and Stol, M. (2004). *Magic and Rationality in Ancient Near Eastern and Graeco-Roman Medicine*. Brill, Leiden/Boston.

Jouanna, J. (1974). *Pour une archéologie de l'école de Cnide*. Les Belles Lettres, Paris.

Jouanna, J. (2017). *Hippocrate*. Les Belles Lettres, Paris.

Laks, A. and Most, G. (2016). *Early Greek Philosophy*. Loeb Classical Library, Cambridge, 524–532.

Lewis, O. (2017). *Praxagoras of Cos on Arteries, Pulse and Pneuma. Fragments and Interpretation*. Brill, Leiden/Boston.

Mansfeld, J. and Runia, D. (1996). *Aëtiana. The Method and Intellectual Context of a Doxographer*, volume 1, Brill, Leiden/Boston.

Nutton, V. (2024). *Ancient Medicine*, 3rd edition. Routledge, London/New York.

Pormann, P.E. (2018). *The Cambridge Companion to Hippocrates*. Cambridge University Press, Cambridge.

Roselli, A. (2018). Nosology. In *The Cambridge Companion to Hippocrates*, Pormann, P.E. (ed.). Cambridge University Press, Cambridge, 180–199.

Roselli, A. (2019). I nomi delle malattie: da Ippocrate a Galeno. In *I nomi del male e i segni dell'eredità. Pensare, nominare e curare la malattia "genetica" dai Greci a noi*, Capocci, M., Cilione, M., Giorgianni, F. (eds). Il Mulino, Bologna, 147–170.

von Staden, H. (1989). *Herophilus. The Art of Medicine in Early Alexandria*. Cambridge University Press, Cambridge.

8

Anatomy and Knowledge of the Human Body: Nerves, Brain, Heart

Vito LORUSSO[1,2]

[1]*Liceo classico F. De Sanctis, Sant'Angelo dei Lombardi, Italy*

[2]*Università degli Studi di Napoli L'Orientale, Naples, Italy*

8.1. Introduction: palpation, dissection and observation

Ψηλαφάω/pselapháo ("palpando quaero vel cognosco aliquem"), meaning "to feel", germ. "betasten", is a word that was used among ancient Greek physicians. Sensation somewhat guarantees corporeality: bodily things are material, palpable and are, in some cases, immediately apparent. Touch is also necessary for the existence and survival of living beings, as noted by Aristotle in a passage of his book *On the Soul* (III 12, 434b, 9–25). It is through anatomical research (*dià tôn anatomôn*) that the peripatetic Alexander of Aphrodisias, a contemporary of Galen of Pergamon, stated that "the heart is the principle of touch and taste" in his *De anima*.

Touch is used by physicians and anatomists to discover and acquire knowledge (*tentare venas* in Latin). In a passage from his commentary on Hippocrates' *On the Nature of Man* (CMG V 9.1, p. 76, 4–6), Galen asks: "Since the compiler of this section on the anatomy of veins cannot see what the one who is palpating/touching might already feel, how can it be said that the same author (*Hippocrates*) compiled both works (in other words, *On the Nature of Man* and the second book of *Epidemics*)?". Practicing touch takes time, but it is fundamental to the art of the

History of Life Sciences and Medicine 1,
coordinated by Lorenzo PERILLI. © ISTE Ltd 2026.

pulse, which Galen himself wanted to learn from the moment he began attending medical school as a boy. Observation can indeed sometimes be achieved through touch: in the case of an internal illness, the physician first palpates the *hypochondria*; in other words, the liver and the spleen, to determine whether these areas are painful or not (diagnosis) but also to deduce that which is palpable/visible from that which is impalpable/invisible. Touch makes it possible to obtain the quality, quantity, tone and force of things. However, the doctor can also touch what the body rejects: "The doctor witnesses frightening displays and touches repulsive things", as we read in the introduction to the Hippocratic treatise *Breaths*.

Palpation can be beneficial to life, according to Aristotle's *History of Animals* (VI 2, 560a, 5–9):

> Eggs that some people call 'sterile' mostly appear in summer. Some call sterile eggs "wind-egg", because it is in spring that female birds take on the warm winds; the same effect also occurs when females are palpated by hand in a certain way.

The doctor palpates during the diagnostic examination, and the patient can also feel their own body, as if to signify the intensity of the pathological condition. In this sense, one could even say that only those who palpate inhabit the practice of care, which is also a relationship of collaboration between caregiver and patient – something that, together with the pursuit of what is beneficial for the sick person and not for the physician, was among the most significant innovations of Hippocratic medicine compared to the previous therapeutic practice linked to sanctuaries and the cult of Asclepius (Figure 8.1).

Figure 8.1. *Marble votive relief, Incubatio, 4th century BCE, Athens, Archaeological Museum of Piraeus*

The clinical record of a woman from Thasos suffering from acute fever, accompanied by chills, mentions palpation as one of the symptoms: “always silent; she palpated, plucked, scratched, picked up hair; tears and, conversely, laughter; she did not sleep...” (*Epidemics* III 17, 15, p. 110 Jouanna). Similarly, the doctor examining her (the Hippocratic physician treats men as well as women) palpates her to judge the fever: “fevers, touching with the gentle hand; cold extremities” (ibid.).

Needless to say, for the anatomical description of living beings and the study of their physiology, dissection (*anatomé*) is actually more incisive than palpation. In this case, it is not just a matter of observing (*theoreîn*) but also of gaining a clearer understanding of the body and its operating mechanisms by breaking it down into its constituent parts. In one of the treatises that make up the collection of *Minor Treatises on Natural History*, or *Parva Naturalia* (*On Breathing* 22, 478a 26–28), Aristotle writes: “To understand how the heart communicates with the lung, we must observe (*theoreîn*) the results of *Anatomical Tables* and *Research on Animals*”. Aristotle stated that breathing, in living beings with a heart and a lung, serves to cool the soul, which is said to be located in the heart and to be inflamed. In the case of fish, on the other hand, which have a heart but no lung, this cooling would take place through the water via the gills. The end of his passage reads as follows: “The position of the heart in relation to the gills must be observed (*theoreîn*) in *Anatomical Tables* and studied in detail in *Research on Animals*”.

Some naturalists and physicians have come to a different conclusion while empirically observing parts of the body, and considered their knowledge to be unsatisfactory:

> We do not know our own bodies, we are ignorant of the positions of their parts and their several functions; and accordingly the doctors themselves, being concerned to know the structure of the body, have cut it open to bring its organs into view, yet nevertheless the empiric school assert that this has not increased our knowledge of them, because it is possibly the case that when exposed and uncovered they change their character (Translation H. Rackham) (Cicero, *Academica* II 122).

8.2. The early days of anatomy

Anatomical investigations of the bodies began very early. Indeed, it could have even been born with man, the very day issues related to surgical diseases and their operations started to surface:

> What could be more useful to a physician for the treatment of war-wounds, for extraction of missiles, for excision of bones, for

> [treatment of] dislocations, fractures with ulcerations [...] than to know accurately all the parts of the arms and legs, and all, not so much of the internal as of the external parts of the shoulders and back, breast and ribs, abdomen, neck, and head? For it is from these that we have to extract weapons, incising the contiguous areas, excising some parts, evacuating humours in putrid infections and abscesses, and treating ulcers. Again with bones we have to cut out some affected parts or open them up. If a man is ignorant of the position of a vital nerve, muscle, artery, or important vein, he is more likely to maim his patients or to destroy rather than save life. Certain knowledge, as the number and appearance of the muscles of the tongue, would be additional, but not primary or essential. (Galen, *On Anatomical Procedures* II 2, pp. 179–181 Garofalo. Translation C. Singer).

From this perspective, anatomical knowledge is primarily justified by its usefulness as an aid to surgery: the physician who possesses the latter, at the very least, does no harm to the patient during the operation. However, wondering why a certain part has certain characteristics, or, more generally, studying the human body almost as an object of philosophy, is by no means time wasted, given that, as Galen himself states just before the passage quoted above, philosophers too have taken an interest in anatomy and not just physicians (II 1, p. 175 G.).

The most recognized example among philosophers is most likely Plato's *Timaeus*, which addresses the subject of the nature of man, placing it within the broader context of the genesis of the Universe: in the prologue to the dialog (27a 3-6), introducing the outline of the speeches yet to be made, Critias says "it seemed appropriate to us that Timaeus speaks as first, given that among us, he is the biggest expert in astronomy and has made the nature of the universe his main focus, beginning with the generation of the universe, and ending with the consideration of the nature of man". However, the perspective of physiology in *Timaeus* is not biological but rather ethical and psychological. The same applies to the description of body parts, with the approach not being anatomical but philosophical/metaphysical, since it is essentially a question of *why*, in other words, the purpose for which the various parts of the body are made as they are. The heart, for example, is said to be "the node of the vessels and the source of the blood that flows with impetus through all the limbs" and is the "headquarters" in relation to the rest of the body (70b 1–3). The brain, on the other hand, which is part of the medulla shaped by the demiurge – the divine intellect that organizes the cosmos for the good – is a round form that "contains within itself, like a plowed field, the divine seed" (73c 6–d 1). The "type of nerves" were conceived by the demiurge as giving the body the ability to bend and stretch, to tense and relax around the vertebrae (74b 3–7). These words already make it clear that, in this case, "nerves" refer to the tendons of modern medicine.

In general, the medical discourse in *Timaeus* focuses on the psychosomatic plexus and the location of the different species of the soul within it: reason is located in the head, separated and protected from the rest of the body by the isthmus of the neck; spiritedness is in the upper part of the thorax, near the heart, separated and simultaneously connected to the divine component (reason), so that it can receive its orders, as if coming from the "acropolis"; last, desire is located in the lower part of the thorax, between the diaphragm and the umbilicus, as far away as possible from the seat of deliberations, where reason resides (69d 6–71d 4).

8.3. Anatomy in the *Hippocratic Corpus*

8.3.1. *Presentation of the material*

Among the 60 or so medical writings in the Ionian language that make up the so-called *Hippocratic Corpus* (a collection handed down from Hippocrates of Cos, homogeneous in language and in the spirit of medical science but heterogeneous in lexicon and doctrinal contradictions), we mention the following because of their more specific focus on anatomy:

– *Anatomy*: this contains a brief description of the trachea, lung, heart, liver, kidneys, bladder and digestive system.

– *Heart*: the description of this organ was carried out by the author by means of dissection (in Chapter 7, the technical verb "to anatomize, to dissect" (*anatémnein*) is repeated, but it is likely only animal dissection, since in Chapter 11, the slain animal is mentioned generically. The physician notices the triangular form of the heart, pericardium and fluid it contains (c. 1), the ventricles separated by a wall (c. 4), the atria (c. 8), the sigmoid valves (c. 10); they also know that "the heart is a very strong muscle, not because of the tendinous parts, but because of the envelope of flesh" (c. 4)). The precision of anatomy is matched by an imaginative physiology: the author has observed the heartbeat but believes that part of an ingested beverage passes through the lung and is sucked up by the heart.

– *Places in Man*: the text opens with a statement about the mutual strength of all the parts of the body:

> Now in my opinion, there is no beginning point of the body, but rather every part is at the same time both beginning and end, in the same way that in the figure of a circle, no beginning point is to be found. (c. 1. Translation P. Potter).

The next chapter begins by describing the nature of the body (cc. 2–8). The opening sentence of Chapter 2 reads: "The nature of the body is the beginning point of medical explorations". The head is thus described, starting with the ears ("the head first has perforations through which things can be heard"), then vessels, tendons (here, as in Plato's *Timaeus*, called "nerves", but this refers to the fibrous tissue "that clasps the joints and extends throughout the body [...] except on the face and in the head, where nerve fibers are found instead" [c. 5]), the bones, the joints, the stomach and the bladder. In the third part (cc. 9–22), flux theory is presented. In a passage from his treatise *On the Natural Faculties*, Galen takes up this Hippocratic theory on the correlation between the body and the spleen:

> Hippocrates says that 'those who have a thriving body have a consuming spleen', and experienced physicians all agree on this point. Conversely, in those with a large spleen, which causes swelling because it is internally purulent, this ultimately destroys the body and makes them full of bad humors… (II 9).

– *Nature of Bones*: this title corresponds to the content of Chapter 1 alone, which mentions the number of bones in the various parts of the body: "There are twenty-seven bones in the hand, twenty-four in the foot, seven in the neck up to the great vertebra…". For the remainder, the treatise consists of a series of descriptions of blood vessels from different sources: some from Aristotle and others from other Hippocratic writings.

– *Anatomy* and *Heart* are recent writings: the former dates from the Hellenistic period (323–31 BCE) or even the Imperial period, and the latter is definitely from the Hellenistic period, with its author being doctrinally close to Diocles of Carystus. *Nature of Bones* could also be recent, although this writing preserves ancient material. The same goes for *Places in Man*. For the latter, we note the presence of the term "esophagus", otherwise only attested in *Anatomy* – here, the term generally designates the entire tract from the oral cavity to the abdomen, although the term "stomach" is also used to describe the cavity close to the viscera in charge of digestion.

8.3.2. *Sources, knowledge and limits of Hippocratic physicians*

Reconstructing the inside of the human body remained a major obstacle for Hippocratic physicians. "It is not possible for anyone to know any of the internal parts of the body (veins, nerves, joints, bones, etc.) if it is observed with the naked eye", notes the author of *On the Art* (Chapter 11), who thus draws a distinction between diseases visible at the skin's surface and internal diseases, which he describes as invisible (they cannot be seen by the eye or by touch), but not invincible: "it is possible to overcome them insofar as the patient allows themselves to be examined and the examiner is willing to scrutinize. [...] What cannot be seen by eyesight is grasped by the sight of reason". In these words, it is clear how Hippocratic physicians, without practicing direct observation of the human body by dissection, tried to get around the obstacle of that which could not be grasped by means of deduction from what could be seen, that is to say, through intelligence.

Although human dissection was not practiced by Hippocratic physicians – nor is it mentioned by the authors of the extant writings – in their day, knowledge of the internal anatomy of animals could supposedly have been nurtured by religious sacrifices, involving cutting the victim open (the fact that religious animal sacrifices were a good opportunity that was within the reach of many to look inside the victims' bodies is also confirmed by Galen in a passage of *On the Therapeutic Method* II 6 (X 121, 3–4 Kühn)). Similarly, hieroscopy or the examination of consecrated victims could foster the development of primitive pathological anatomy. In general, Hippocratic physicians are rather discreet about the sources of their knowledge of the internal structure of human body parts. The author of *On the Nature of Man*, for example, gives a detailed description of blood vessels, without justifying his knowledge in any way, as if it were a given.

Notably, in Galen's commentary on this Hippocratic writing, quoted at the beginning of this chapter, the comparison between the irrelevance of the description of veins in *On the Nature of Man* and the corresponding description in the long section devoted to anatomy in *Epidemics* II, allows Galen to conclude that the only authentic description of the vein system that has been preserved in the writings attributed to Hippocrates is specifically the one in *Epidemics*. Furthermore, the description of the four pairs of veins in Aristotle's *History of Animals* (III 3, 512b 12–513a 7) is derived from Chapter 11 of *On the Nature of Man*, but Aristotle instead attributes it to Polybus, Hippocrates' disciple, son-in-law and the successor as head of the Cos school. This same description also appears in Chapter 9 of *Nature of Bones*. This situation famously demonstrates the extent to which the Hippocratic Corpus is both layered and fluid.

The author of *Sacred Disease* in turn describes the brain ("in man, it is doubled…, a thin membrane divides it in the middle"), its veins ("veins come from all over the body to the brain, numerous and fine; two, however, are large: one from the liver, the other from the spleen…", c. 3), and in Chapter 11, he gives evidence that he has performed a dissection of the brain ("if you open the head, you will find that the brain is moist, full of an aqueous and evil-smelling humor…"). He does not, however, disclose his sources. Perhaps the fact that the transmission of knowledge within the medical families of the Asclepiadae of Cos and Cnidus was done verbally and exclusively to the male line could explain the Hippocratic authors' silence. Galen, in *On Anatomical Procedures* (II 1, p. 175 G.), reminds us that having learned to dissect bodies as a child was, for the Ancients, an absolute guarantee that the method would not be forgotten as an adult.

Hippocratic physicians were therefore knowledgeable about the main organs of the human body (brain, heart, lungs, liver, kidneys, spleen and bladder), even if they did not call them "organs" – just "parts". The terms "nerve", "vein" and "artery" have a different meaning for us in the contemporary era: nerves, as we have already mentioned, are actually tendons, in other words, ligaments that hold the body together and make it possible to generate movements by greater or lesser tension; veins are, generically, blood vessels and are hardly ever distinguished from arteries; and the "artery" is actually the trachea. A good knowledge of the human skeleton is demonstrated in surgical treatises of the *Hippocratic Corpus*: perfect knowledge of the shape of bones and ligaments is necessary to correctly diagnose and treat dislocations and fractures. One particularly serious, and therefore incurable, affliction for the author of *Articulations* is the inward dislocation of vertebrae:

> There are even some who think this is the easiest distortion to recover from, not even requiring reduction, but that such injuries get well of themselves. There are many ignorant practitioners ; and they profit by their ignorance, for they get credit with their neighbours… They think that the projecting ridge along the spine represents the vertebrae themselves… (c. 46 Translation E.T. Withington).

These words show how much anatomical knowledge could vary from one physician to another, particularly regarding the way in which medical art was taught. The most educated physicians were fully aware of this and were able to specifically reason about the invisible world of the interior of the human body.

8.4. Anatomy according to Aristotle

8.4.1. *Writings, context, working methods, data organization*

Among Aristotle's works devoted to biology, and consequently also to the anatomical description of living beings, we find the following: *History of Animals* (over 500 different species are described, most of them observed during the philosopher's stay in Asso and Mytilene between 347 and 342 BCE); *Parts of Animals* (in which animals are classified according to their habitat and anatomical characteristics); *Movement of Animals* (on the anatomophysiology of the organs responsible for movement); *Generation of Animals*, which illustrates the function of reproduction; *Parva Naturalia – Minor Treatises on Natural History*, a collection of monographic booklets on psychological or physiological subjects; *On the Soul* (in three books, dealing with the soul from a biological perspective, as what distinguishes a living body from a dead one, thus embracing the entire domain of living beings: in the opening lines of the treatise, Aristotle writes that "the soul is almost a principle of living beings" [I 1, 402a 6-7]); *On the Movement of Animals* (the specific subject of the work is the movement of animals in its broadest sense, involving both change and motion).

A priori, it is not impossible to imagine that Aristotle – whose father was an Asclepiad, i.e. a member of a family that claimed descent from the medicine god Asclepius, passing on medical knowledge from father to son – was in a sense the direct heir to the physicians of the Hippocratic Corpus. The collaboration and the difference between the naturalist and the physician are recalled several times in his writings, for example, in *Sense and Sensibilia* (c. 1, 436a 17–b 1), *On the Soul* (I 1, 403a 27–b 19), *On Breathing*:

> Health and disease also claim the attention of the scientist, and not merely of the physician, in so far as an account of their causes is concerned. The extent to which these two differ and investigate diverse provinces must not escape us [...] Physicians [...] make some mention of natural science, and claim to derive their principles from it, while the most accomplished investigators into nature generally push their studies so far as to conclude with an account of medical principles. (c. 27, 480b 22–30. Translation G.R.T. Ross).

In a beautiful page from *Parts of Animals* (I 5, 645a 5–17), the study of living beings is elevated to a philosophical discipline status, as long as the scientist knows the causes of facts:

> It remains to speak about animal nature, omitting nothing in our power, whether of lesser or greater esteem. For even in the study of animals disagreeable to perception, the nature that crafted them likewise provides extraordinary pleasures to those who are able to know their causes and are by nature philosophers. Surely it would be unreasonable, even absurd, for us to enjoy studying likenesses of animals - on the ground that we are at the same time studying the art, such as painting or sculpture, that made them - while not prizing even more the study of things constituted by nature, at least when we can behold their causes. For this reason we should not be childishly disgusted at the examination of the less valuable animals. For in all natural things there is something marvelous (Translation J. Lennox).

Aristotle's direct knowledge of Hippocratic writings such as *Nutriment*, *Places in Man*, *Diseases 4*, *Sacred Disease*, *On the Nature of Man*, and *Breaths* is plausible.

In fact, in *History of Animals* (III 2–3, 511b 1–513a 7), Aristotle reports and refutes the opinions on the vein system of the physician Syennesis of Cyprus, Diogenes of Apollonia, the aforementioned Polybus and other physicians not specifically named but who may belong to Hippocrates' group: "there are some writers on Natural History who have not ventured to lay down the law in such precise terms as regards the veins, but who all alike agree in assigning the head and the brain as the starting-point of the veins. And in this opinion they are mistaken" (513a 8–12. Translation D'A. Thompson). Aristotle is rather indulgent toward these physicians, given the objective difficulty of making observations on the veins of dead and living animals (511b 13–23). Thus, Aristotle may have acquired some of these anatomical details on animals by sharing information with hunters, fishermen, cooks and physicians.

Aristotle was able to describe the natural arrangement of veins and establish their origin from the heart owing to a new method of observation he advocated, which involved killing animals by suffocating them after they had lost weight (513a 12–15). This is not Aristotle's only contribution to research in the field of biology and, more specifically, medicine. Indeed, the comparative anatomy between man and other living beings, which seemed similar to Aristotle, enabled him to derive information about the interior of the human body and how it functions:

> To begin with, we must take into consideration the parts of Man. For, just as each nation is wont to reckon by that monetary standard with which it is most familiar, so must we do in other matters. And, of course, man is the animal with which we are all of us the most familiar. Now the parts are obvious enough to physical perception. However, with the view of observing due order and sequence and of combining rational notions with physical perception, we shall proceed to enumerate the parts: firstly, the organic, and afterwards the simple or non-composite (*History of Animals* I 6 (491a 19-26. Translation D'A. Thompson)).

In these words, we can see the working method described at the beginning of *Physics* (I 1, 184a 10–18): moving from what is best known and most obvious to the observer, to reach that which is clearest and most familiar by its very nature. Research in human anatomy is achieved in the same way as research in biology and zoology. The aims of these three disciplines are to describe living beings, in order to explain their material cause, and illustrate their specific shapes and purposes. From this, we can perhaps better understand why Aristotle's object of interest was not so much the animal as a whole but its individual parts.

In the history of biology and medicine, in animal anatomy, Aristotle introduced the distinction between homeomerous, in other words, homogeneous and non-composite parts (e.g. meat, bones, tendons and veins but also fluid or soft substances such as blood, sebum, sperm, bile and milk) and non-homeomerous or composite parts ("for example, the hand cannot be divided into hands, nor the face into faces"), a distinction that can be found as early as the beginning of the *History of Animals* (I 1, 486a 5–14). The internal body organs, such as the heart, lungs and liver, also belong to the group of non-homogeneous parts. The viscera and heart, being both sensory organs and instrumental parts, represent a complex that can be divided into homogeneous parts but is not itself homogeneous in shape and configuration (647a 30–33). The terms "organ" and "organic" are applied to each part of the body when it serves as an instrument for the functioning of the body as a whole. Serviceability and functionality are therefore the criteria by which nature has attributed organic/instrumental parts to living beings. This last point is clear when we consider why human beings have hands:

> Nature has provided the lesser to the greater and superior, not the more honourable and greater to the lesser. So [...] it is not because they have hands that human beings are most intelligent, but because they are the most intelligent of animals that they have hands. For the most intelligent animal would use the greatest number of instruments well, and the hand would seem to be not one instrument, but many; indeed it is, as it were, an instrument for instruments (*Parts of Animals* IV 10, 687a 13–21 Translation J. Lennox).

In terms of the final cause (teleology), Aristotle justifies the presence of organs and other parts in the body as a function of the soul and the latter's control over the body. A passage in *Parts of Animals* (IV 10, 686a 5–31) explains that the head is above all a function of the brain and that its position in the upper region of the body in relation to that of the heart and stomach, as well as the fact that it is the area of the body in which the eyes are located and where there is an orifice for food to enter the body, all depend on this.

8.4.2. *Heart and brain*

For Aristotle, the heart is the body's main organ of control. "It is the origin of the blood vessels [...] and its nature is vascular as if it were like in kind to them." (*Parts of Animals* III 4, 665b 15–17). The heart's central position in the human body also corresponds to its primary, hegemonic role. The body's other vital center is the brain.

An anatomical description of the brain can be found in *History of Animals* I 16 (494b 24–495a 11): the brain is located in the front part of the head; it is surrounded by two membranes (the one surrounding the brain, with many veins, is called *meninx* by Aristotle). This is made up of two parts: it is always devoid of blood (same observation in Hippocrates' *Regimen* II 49) and veins inside, it is naturally cold to the touch and has a ventricle in the middle. Aristotle also mentioned the cerebellum (*parenkephalís*), "located at the posterior end of the brain, differing from it in shape as we may both feel by touch and see" (494b 33). In humans, of the same size as other animals, the brain is larger and more moist. The functions and relationships of the brain to the heart are described in *Parts of Animals* II 7 (652a 24–653b 8): following from Plato's *Timaeus*, Aristotle emphasizes the connection between the brain and the spinal cord, even though the former is cold by nature, and the latter is warm. However, through an autopsy, he excludes the brain's connection with the sensory parts and explains its presence in animals as a function of the preservation of the whole organism; in other words, as a "counterweight" to the heat coming from the heart. "And it is for this reason that animals, whose nature consists of a combination of water and earth, possess a brain". Both are indeed cold elements, but water is qualitatively fluid, whereas the earth is solid, which helps us understand the particular consistency of the brain (Figure 8.2).

Figure 8.2. *Hieronymus Bosch, The Lithotomy, or The Extraction of the Stone of Madness (1490), Madrid, Museo del Prado. For a color version of this figure, see www.iste.co.uk/perilli/history1.zip*

Lastly, the coldness of the brain would also explain why flows in the body originate in the head: this would occur through the cooling of hot flows from the parts below the head. Thus, for Aristotle, the cooling of the whole body explains the role of the brain. However, in the body's overall efficiency, the brain's role remains secondary to that of the heart. It is worth noting that while in *Parts of Animals* the brain is not recognized as the center of the faculties of perception and intelligence, in a passage of the *History of Animals* the connection between the brain and the optic nerves is identified, and consequently, Aristotle seems to accept the encephalocentric thesis on perception that had also been supported by Alcmaeon, Hippocrates and Plato, among others:

> From the eye there go three ducts to the brain: the largest and the medium-sized to the cerebellum, the least to the brain itself; and the least is the one situated nearest to the nostril (I 16, 495a 11–14. Translation D'A. Thompson).

The heart, conversely, has a perceptive function for Aristotle. As previously mentioned, it is the principle of the veins and is venous in nature. In man, as in other animals, the heart occupies "the position [...] that befits a principle: [...] in the center, more up than down and more forward than back [...] in the center, therefore, of the necessary part of the body". As a part and principle of the veins, the heart is not crossed by any of them; its main zone is a dense, hollow component filled with blood. Aristotle knew of three heart cavities: the two ventricles and the left atrium. On the basis of new dissections and embryological research, Aristotle established that the heart is also "the principle or source of blood and its first receptacle". The heart is also the foundation of body mobility and therefore possesses an abundance of tendons (*Parts of Animals* III 4, 665b 9–667b 14). We might wonder whether these tendons (tendinous cords?) observed by Aristotle around the heart are the same as the "ribs of the soul" mentioned by Plato in a passage from the third book of the *Republic* (411a 5–b 4), in which he warns men against the risk of excessively indulging in the sweetness of music. According to the latter, music has the power to sever the ribs after causing the soul to melt.

8.5. Anatomy according to Diocles of Carystus and Praxagoras of Cos

Around the middle of the fourth century BCE, we come across Diocles of Carystus, whom the Athenians called "the new Hippocrates" and whom Aristotle may have even known himself. According to Galen (*On Anatomical Procedures* II 1, p. 177 G.), Diocles was the first to write about anatomy, although it is not clear whether he was writing about human anatomy or animal anatomy more generally. What is certain is that, from the fourth century onward, dissection gradually separated from pathology; in other words, it no longer aimed to corroborate an opinion about disease – as was still the case with Hippocratic physicians – but rather to describe the human body, despite the revulsion that research of this kind could bring about in contemporaries. In a passage from *Parts of Animals*, Aristotle mentioned that "it is with great disgust that we see what man is composed of: blood, flesh, bones, veins and other parts of this kind" (I 5, 645a 28–30). However, Diocles' achievements in this field were far more unassuming than those of later Alexandrian anatomists. In a stark judgment of Diocles and the empirical physicians, worshippers of so-called casual anatomy, Galen states that "the usefulness of what is written is not explicated, but everything is thrown in there at the same level, both what is of most use for the art and what is of little or no use". Around the same time as Diocles, Praxagoras of Cos also showed a strong interest in animal anatomy, conducting research on the brain and spinal cord in particular. However, our knowledge of these two physicians is heavily conditioned by the fact that we only possess fragments of their writings from indirect testimonies, with all that this entails in terms of both the evaluation of sources and historiographical reconstruction.

8.6. Anatomy according to the Alexandrians: Herophilus and Erasistratus

The 'invention' of human anatomy was the essential novelty of Hellenistic medicine, practiced in Alexandria by Herophilus of Chalcedon (approximately 335–280 BCE), a pupil of Praxagoras, and Erasistratus, born in Ioulis on the island of Ceos, a pupil of the physician–philosopher Chrysippus of Cnidus, who was active approximately 250 BCE. The context in which these two physicians worked for most of their lives was the Alexandrian Museum-Library, an institution that aimed to collect literary and scientific documents from libraries around the world, encouraging research. This scenario may explain the critical interest in Hippocratic writings from Herophilus and his pupils.

In the case of Herophilus and Erasistratus, we only have secondhand knowledge; in other words, information and fragments of their writings have been passed on to us from other ancient physicians. Herophilus was interested in anatomy and practiced the dissection of human cadavers. He adhered to the following principle: "Describe phenomena first, even if they were not first". Although he was probably not one of the physicians of the Ptolemaic court, he was likely connected to political power, being the first to not only systematically practice the dissection of human cadavers but also seemingly perform vivisection on convicts in the king's prisons. This did not cause any scandal, as Aulus Cornelius Celsus, an early imperial Roman physician, informs us (*incidere corpora mortuorum*, "dissecting corpses"; *nocentes homines vivos incidere*, "to dissect criminals still alive", Celsus, *Medicine*, proem 23–26 = text 63a, p. 187 von Staden). Herophilus specifically studied the nervous system, vascular anatomy, heart and brain anatomy, respiratory system, reproductive system and eye. He is credited with describing the torcular, the endocranial venous vessel located just beside the internal occipital protuberance: "A sort of cavity, almost a receptacle, which Herophilus used to call a torch for this very reason. From it, as from an acropolis, run many channels to the lower parts of the body", writes Galen in *On the Usefulness of the Parts of the Body* (IX 6 = fr. 123 von Staden). Herophilus is also credited with the discovery of the different rhythms of the pulse described according to the different ages of patients (Pliny, *Natural History* XXIX 6), as well as the invention of the technical terms *arachnoeidés* ("shaped like a spider-web") for the retina and *dyodekadáktylon* ("twelve-fingered") for the duodenum. Erasistratus studied the heart valves, intestines, kidneys and nervous system. He is credited with discovering the chyle cistern and abdominal swelling in patients who die of ascites.

Within the Alexandrian school of medicine, the rational, experimental approach to knowledge of the entire human body greatly favored the development of anatomy and physiology, and thus of pathology and therapy. The most significant discovery of Herophilus and Erasistratus is undoubtedly related to the nervous system. The Greek word used by these two physicians for "nerve" is *néuron*, originally used interchangeably for tendons and ligaments. Before Herophilus and Erasistratus, the sensory and motor nerves, as well as the mechanisms for transmitting sensation and controlling muscles, were not known. Needless to say, with the development of knowledge about the nervous system, the location of the human soul (*hegemonikón*) shifted from the heart to the brain.

The school of the followers of Erasistratus (the Erasistrateans) and Herophilus (the Herophileans) flourished until the Imperial era. Erasistratus and Herophilus were also significantly active outside Alexandria, from the second half of the 2nd century BCE. In fact, the expulsion of physicians and scholars from Alexandria ordered by Ptolemy VIII Euergetes dates back to 145–144 BCE: an action that had the opposite effect of spreading men, ideas and books to other centers of the Hellenistic and Roman world.

8.7. Anatomy in the early imperial era

An important but troublesome text of medical doxography, probably dating from the 1st century CE, which was conveyed through a papyrus of the same period (the so-called Anonymous of London) and was dependent on studies of the history of medicine carried out in the 4th century BCE by a disciple of Aristotle, not only describes the structure and operating mechanisms of the human body but also critically discusses Herophilus' and Erasistratus' perspectives on this subject.

For example, the London papyrus states that blood is naturally contained in the arteries, refuting Erasistratus' opinion that arteries only contain *pneuma*. The papyrus also refutes Herophilus' opinion: the distribution of food between body parts does not take place through the arteries but through the veins.

Another anatomist of the late 1st century CE is Rufus of Ephesus, the author of an anatomical lexicon (the *Onomasticon*). This is an isagogical (i.e. introductory) work, which describes the parts of the human body with the appropriate terminology, organizing the subject from head to toe. The first section of the lexicon describes the external parts, and the rest addresses the internal parts. In the first part, the anatomy is that of a slave (*páis*) serving as an 'object of demonstration'. In the second, however, Rufus refers to the ape (*píthekos*) as being the most anatomically similar to humans of all living creatures. In the educational practice of the time, Rufus' lexicon enjoyed good fortune and served as the basis for two reworkings, one of which was more specifically dedicated to the anatomy of the skeleton, preserved in the subsequent Greek manuscript tradition under the name of Rufus himself. The act of redoing/adapting/correcting medical writings intensified, especially after Galen's great achievements, but Rufus' material can certainly be identified in medical compilations and encyclopedias of late antiquity and the early Byzantine period.

8.8. Anatomy according to Galen

8.8.1. *Usefulness of anatomical investigation*

The significant work of Galen of Pergamon (129–c. 215 CE) not only encompasses all branches of medicine but also logic, philosophy and philology. Much of it is also autobiographical, so both his character and his career are well known. Galen received excellent medical education and, even before that, literary and philosophical education from his father Nikon, a wealthy and highly cultured architect. He first began studying medicine in Pergamon and then in Smyrna under various teachers, mainly Satyros, a follower of an important school of anatomy, and Pelops, a Hippocratic physician. He then moved to Corinth to attend the lectures of the great anatomist Numisianos, who had also mentored Pelops. When Pelops moved to Alexandria, which was still a large center for medical studies, particularly anatomy, Galen decided to follow him. His stay in Alexandria lasted approximately five years, during which time Galen was able to see and listen to many excellent mentors, some of whom, combining experience and practice, were able to train their pupils using Hippocratic techniques, much like the famous passage from Plato's *Phaedrus* (270c 1–d 7); some other mentors were physician–sophists, and others, lastly, were truly bad mentors, such by Galen as Julian, who, in a well-known passage in the first book *On the Therapeutic Method*, was presented to us by Galen as a phony. Upon completing his studies and returning home, Galen's professional life began when he was first appointed as a physician to the gladiators of Pergamon. Then, in 162, he went to Rome for the first time, where he stayed for three years. In addition to his work as a physician to members of the ruling class, he also devoted himself to anatomical research. His second and longest stay in Rome, which began around 168 and took place at the imperial court as personal physician to Commodus, the son of Emperor Marcus Aurelius, was also a period of intense scientific activity for Galen. Maintaining a balance between innovation and tradition, Galen was a fundamental successor of the great Hippocratic and Hellenistic medical tradition. Although he was a student without a school and a mentor without disciples, Galen felt the continuity of the medical teachings of Hippocrates as effectively in operation throughout his long professional life, strongly convinced of the value of early training. As a historian and doxographer, and despite the challenges and distortions he inflicted upon the opinions of his adversaries, Galen is, in many cases, a unique witness to medicine and philosophy's long history, whose knowledge, without his mediation, would be lacking in many essential aspects.

Galen's emphasis on phenomena, in other words, the observation of visible facts, explains the key role of anatomy in his work. In the autobibliographical work *On My Own Books*, he reviews (Chapter 3) all his writings on anatomical doctrine: *On Bones*, *The Dissection of Veins and Arteries*, *The Dissection of Nerves*, *The Dissection of Muscles*, *On Anatomical Procedures*, the latter in 15 books. Of all these anatomical writings, the first four are aimed at beginners. For educational purposes, we also know that Galen wrote epitomes of books by previous illustrious anatomists, particularly Marinus, one of his mentors and the author of *Anatomy*, a 20-book series that Galen summarized in a new work in four books. Marinus was a custodian of Alexandrian anatomical knowledge and discoveries, particularly in the field of brain anatomy. In fact, Galen writes of Marinus' *Epitome*:

> In the seventeenth book, he (Marinus) speaks of the brain's prominence. In the eighteenth, he looks at voluntary functions and specific nerve differences [...] In the nineteenth, he speaks of the nerves that come from the brain, [...] the origin of the sense of smell and the nerves that go to the eyes [...].

In this transmission of Alexandrian knowledge, Galen was also justifiably proud of his anatomical knowledge of the brain and nervous system. While muscles are responsible for the movement of different parts of the body, he writes, they also need nerves from the brain to move, as demonstrated by voluntary or accidental amputations of muscles and nerves. In turn, nerves can perform their tasks and interact with the brain through what is known as *pneuma*, the vital force determining physiological and sensory processes in the body. While the first physicians and naturalists to speak of *pneuma* or spirit, an ethereal substance, considered its origin to be the arteries (for example, Aristotle and Erasistratus), Galen thought that the psychic *pneuma*, responsible in living beings for consciousness, sensitivity and voluntary movement, was contained in the brain–nerve system. This *pneuma* is produced from arterial blood, which is refined in the so-called *rete mirabile*, a vascular system at the base of the brain, before being circulated to the cerebral ventricles and then to the nerves. The *pneuma* carries impulses from the brain to parts of the body and sensory information from the organs to the brain; this is what, for example, acoustic and optic *pneuma* do. Another significant discovery of Galen's in the field of nerves was the recurrent laryngeal nerve.

In addition to being the main way of understanding physiological processes, for Galen, the usefulness of anatomical knowledge lies in the fact that it can be used to avoid errors in diagnosing diseases and, above all, in surgical interventions:

> When ligaments and tendons are round, they look like nerves to the uninitiated, especially if they are not aware that they are much harder than nerves. Conversely, when ligaments and tendons are flat, they know that they are something other than nerves, although they do not know what. This is because they are unable to distinguish one from the other. However, since you know the nature of each of these three things (ligaments, tendons and nerves), as well as their conformation and position in the body in relation to each part, when you will see an injury in that part of the body, you will be trained to recognize whether it is a nerve, a ligament or a tendon that is injured (*On the Therapeutic Method* VI 4, X 409–410 Kühn).

8.8.2. *Galen the anatomist: new discoveries and acquisitions*

Galen provides us with an account of his own anatomical research in the previously mentioned *On Anatomical Procedures*, a work of 15 books, composed at different stages from approximately 177, whose history was marked, such as so much else in Galen, by the burning of the Roman Temple of Peace in 192 (in this tragic circumstance, the manuscripts of the work were lost; Galen found no other copies of the last four books and therefore had to rewrite them, as he himself tells us). When reading some of the pages of this treatise, it is almost like entering Galen's workplace: there is even mention of the room's luminosity and temperature, allowing us to better imagine the place where Galen worked. Some of the lines at the end of Book XI are particularly profound:

> When I finish the work on which I am now currently working, I shall also attend to the dissection of these animals (i.e., *certain reptiles, fish and birds*) and the description of what can be seen in them; I do this, as you know, while the animal is before me and I observe these things in it, especially when I illustrate the method to be followed in dissecting it (*On Anatomical Procedures* XI 12, p. 949 Garofalo).

Starting from the results obtained by his mentors, Galen's most famous studies concern the neurological and vascular systems, in particular: the description of the four cerebral ventricles and the part that flows into the medulla, "which anatomists have likened to the end [...] of a feather, similar to an awl [...]" (*On Anatomical Procedures* IX 7, p. 829 G.: Galen states that observation is easier in an animal that is "old and thin, afflicted by hunger and fatigue, [...] made small because it is emaciated by disease" and whose brain substance is as hard and dry as possible); the description of the two optic nerves to which a part of the brain is connected, namely the *corpus geniculatum laterale*; the observation of the ducts that lead from the brain to the nostrils, an observation that Galen states he made "several times [...] and in places where there was bright light" (*On Anatomical Procedures* IX 7, p. 831 G.); open-brain experiments to test its effects, including loss of breathing, voice, movement and sensitivity (to make these experiments more successful, Galen recommends removing the thick membrane from the animal's brain in summer or, if this is done at another time of year, in a heated room, *On Anatomical Procedures* IX 12, p. 846 G.); experiments on the effects of compressing the two anterior ventricles of the brain: Galen states here that "if you take an animal and compress a part of the two anterior ventricles, in any part where the root of the nerves of sight is located, the animal ceases to close its eyes, even if you bring something close to its pupils, and the appearance of the eye on the side of the ventricle of the brain that is being compressed becomes similar in appearance to the eyes of a blind man" (*On Anatomical Procedures* IX 12, p. 847 G.); serial sectioning of spinal nerves; some experiments on the nerves and muscles of the thorax responsible for breathing and the voice; sectioning of the carotid artery; insertion of a tube in the femoral artery; and various embryological experiments. All of these experiments were of great significance to Galen, both in terms of the *more geometrico* demonstration of the existence of facts that were not immediately obvious but deducible through the application of the Aristotelian system of causality (the shape and material of a part can explain why something is the way it is), and in terms of promotion: this is indeed the way in which Galen could prove the superiority of his own knowledge over that of rival physicians, notably the hated Methodists. Regarding the latter,

Galen states the following: "One cannot count the number of those who ended up in the Paeonian (ironically for "healing") hands of the Methodists and died in atrocious torments" (*On the Therapeutic Method* (VI 2 [X 390 K.])). After all, technical competence, a polemical temperament and a flair for the spectacular and dramatic were all part of Galen's character, even when he was dealing with anatomy.

A point worth mentioning here is the discrepancies in anatomy, with Galen devoting a two-book work to the subject: *On the Differences between Anatomists*, an introductory work along with the *Anatomy on Cadavers* (in one book) and the *Anatomy on the Living* (in two books) for the study of natural functions, described in the three books of *On the Natural Faculties*, as well as psychic functions, dealt with in several other works. On the basis of advances in knowledge, Galen reduced the differences in anatomy to facts (e.g. the existence of an anatomical structure), names (such as the naming of the layers of the eye) and, last, the way in which the facts themselves were presented (for example, the grouping of various muscles under a single name or several).

8.9. Galen's other works

Despite discoveries in anatomy, even after Galen, the therapeutic impact was rather small. Physicians generally continued to adhere to the ancient Hippocratic pharmacopoeia, but the proponents of empiricism leaned toward discovering the treatment of pathological conditions in an experimental way, namely, through a procedure based on sensitive perception. Anatomy played a key role here. Indeed, it could provide all of the data from which the empirical physician, when diagnosing or treating something less immediately perceptible to them, would take what is known as the "transition to the similar". This was not the case with methodical physicians, disciples of Thessalos of Tralle (in Nero's time, 1st century CE). Galen specifically engaged with Thessalos in repeated polemics, especially since Thessalos had been a critic of the entire category of previous physicians, including Hippocrates. This arrogant attitude is also reflected in the preface to a pharmacology textbook *On the Virtues of Plants*, which is classically attributed to Thessalos and reads as follows: "none of the previous physicians having been able to keep their promises [...], I alone among all men, even generations past, have succeeded in doing something extraordinary...". Furthermore, for Methodists such as Thessalos, anatomy was not essential to the practice of medicine, and this fact alone placed him in radical opposition to Galen, who criticized him for having made the medical career accessible – with no regard for tradition – to aspiring physicians from the most modest social classes by introducing an extremely simplified curriculum and charlatanism.

8.10. Anatomical illustration in manuscripts

To conclude this chapter, let us consider anatomical illustrations. Aristotle refers to anatomical tables (*anatomái*) as an indispensable aid for the study of the living, for example, in *History of Animals* III 1 (509b 22) and *Parts of Animals* III 14 (674b 16): whoever reads Aristotle's treatise will be able to observe (*theoréin*) the position, shape and differences of the parts. However, these tables have been lost. Galen speaks of his own anatomical drawings in a passage that has only been preserved in an Arabic translation of *On Anatomical Procedures* (II 3, p. 187 G.): for Galen, both treatise and drawings are important educational tools that are far superior to occasional anatomy and to anatomy professed from a chair, and are useful for observing wounds and learning about anatomy without direct practical experience. On the basis of Aristotle and Galen's comments, we can only imagine how useful the anatomical drawings in their natural history books were as a mnemonic tool, allowing the reader to momentarily recall the nomenclature, position, structure and function of the body's various organs at a glance. The images had to be clear, sharp and not overwhelmed with detail.

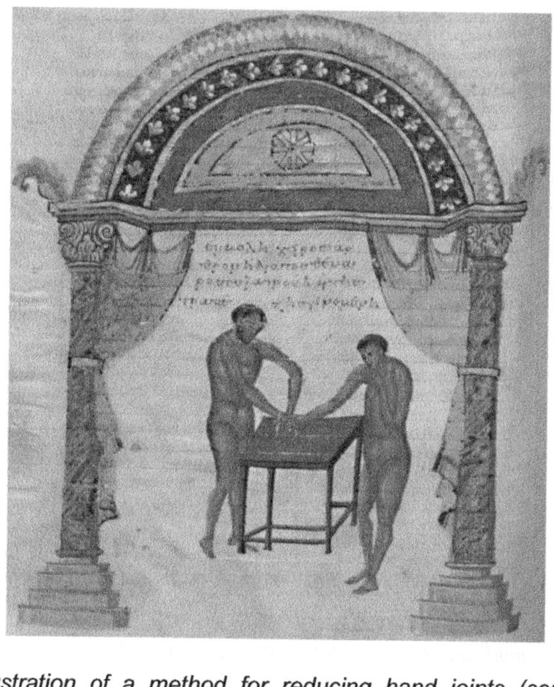

Figure 8.3. *Illustration of a method for reducing hand joints (source: Florence, Biblioteca Medicea Laurenziana, ms Plut. 74.7, f. 196v, with permission from the Ministero della cultura; reproduction is prohibited). For a color version of this figure, see www.iste.co.uk/perilli/history1.zip*

An example of an anatomical–surgical illustration is provided by an 11th century Byzantine manuscript, now preserved in Florence's Biblioteca Medicea Laurenziana, containing, among other things, a commentary by the empirical physician Apollonius of Citium on Hippocrates' *Joints*, from the 1st century BCE, in three books. In the Laurentian manuscript, an illustrative image with a short caption follows the explanation of each joint reduction method (Figure 8.3).

All of these illustrations – despite a certain uniformity of the patients (always depicted naked but asexual), the physicians and their assistants (sometimes depicted naked, sometimes dressed in tunics and cloaks) and the scene of the treatment itself – attest to the extent to which the tradition of naturalistic iconography of the human body was preserved in the Byzantine East. A similar consistency can be observed in manuscripts from the Latin West around the 12th century: there is a series of five figures typically illustrating the veins, arteries, bones, nerves and muscles of the human body, the origins of which date back to the great school of anatomy in Alexandria in the 1st century CE.

8.11. References

English translations of primary sources

Aristotle (1910). *The Works of Aristotle translated into English under the editorship of J.A. Smith and W.D. Ross – vol. IV, Historia Animalium*, Wentworth Thompson, D. (trans.). Clarendon Press, Oxford.

Aristotle (1931). *The Works of Aristotle translated into English under the editorship of J.A. Smith and W.D. Ross – vol. 3, Parva Naturalia*, Beare, J.I. and Ross, G.R.T. (trans.). Clarendon Press, Oxford.

Aristotle (2002). *On the Parts of Animals I-IV*, Lennox, J.G. (trans.). Clarendon Press, Oxford.

Cicero (1967). *De natura deorum; Academica,* Rackham, H. (trans.). Harvard University Press, Cambridge, MA/London.

Galen (1956). *On Anatomical Procedures,* Singer, C. (trans.). Oxford University Press, London.

Hippocrates (1927). *Hippocrates*, volume III, Withington, E.T. (trans.). Heinemann, London.

Hippocrates (1995). *Places in Man. Glands. Fleshes. Prorrhetic 1-2. Physician. Use of Liquids. Ulcers. Haemorrhoids and Fistulas.* Potter, P. (ed.). Harvard University Press, Cambridge/ Mass.

Further studies and translations

Boudon-Millot, V. (2007). *Galien. Tome I : Introduction générale. Sur l'ordre de ses propres livres. Sur ses propres livres. Que l'excellent médecin est aussi philosophe.* Les Belles Lettres, Paris.

Boudon-Millot, V. (2012). *Galien de Pergame. Un médecin grec à Rome.* Les Belles Lettres, Paris.

Brockmann, C. (2011a). Manuskriptanalyse und Edition II: weitere Beispiele aus Galens Kommentar zu Hippokrates, De articulis. In *Officina Hippocratica*, Perilli et al. (eds), 329–344.

Brockmann, C. (2011b). Mediziner. In *Aristoteles-Handbuch*, Rapp et al. (eds), 40–42.

Brockmann, C. (2011c). Organe. In *Aristoteles-Handbuch*, Rapp et al. (eds), 286–292.

Brockmann, C., Brunschön, W., Overwien, O. (eds) (2009). *Antike Medizin im Schnittpunkt von Geistes- und Naturwissenschaften. Internationale Fachtagung aus Anlass des 100-jährigen Bestehens des Akademienvorhabens.* Corpus Medicorum Graecorum/ Latinorum, De Gruyter, Berlin/New York.

Brunschwig, J., Lloyd, G.E.R., Pellegrin, P. (eds) (1996, new edition 2021). *Le savoir grec.* Flammarion, Paris.

Burkert, W. (2009). Die Entdeckung der Nerven. Anatomische Evidenz und Widerstand der Philosophie. In *Antike Medizin*, Brockmann et al. (eds), 31–44.

Byl, S. (1977). Les grands traités biologiques d'Aristote et la Collection hippocratique. In *Corpus Hippocraticum*, Joly, R. (ed.), 313–326.

Cambiano, G. (1977). Le médecin, la main et l'artisan. In *Corpus Hippocraticum*, Joly, R. (ed.), 220–232.

Capitani, U. and Garofalo, I. (1986). *Gaio Plinio Secondo. Storia naturale IV: medicina e farmacologia. Libri 28–32.* Einaudi, Turin.

Debru, A. (2021). Galien. In *Le savoir grec*, Brunschwig, J. et al. (eds), 735–748.

Deichgräber, K. (1930). *Die griechische Empirikerschule: Sammlung der Fragmente und Darstellung der Lehre.* Weidmann, Berlin.

Di Benedetto, V. (1972). La citazione di Polibo nella *Historia Animalium* di Aristotele. In *Studi Classici in Onore di Quintino Cataudella*, volume 2. Catania, 225–233 (reprinted in: Di Benedetto, V. (2007). *Il richiamo del testo: contributi di filologia e letteratura*, volume 4. Edizioni ETS, Pisa, 1727–1733).

Di Benedetto, V. and Lami, A. (1983). *Ippocrate. Testi di medicina greca.* Rizzoli, Milan.

van der Eijk, P. (2000). *Diocles of Carystus: A Collection of the Fragments with Translation and Commentary.* Brill, Leiden/Boston/Cologne.

van der Eijk, P. and Francis, S. (2009). Aristoteles, Aristotelismus und antike Medizin. In *Antike Medizin*, Brockmann et al. (eds), 213–233.

Ernout, A. (2003). *Pline l'Ancien. Histoire naturelle. Livre XXIX.* Les Belles Lettres, Paris.

Ferrari, F. and Petrucci, F.M. (2022). *Platone. Timeo.* Fondazione Lorenzo Valla, Mondadori, Milan.

Flashar, H. (2013). *Aristoteles: Lehrer des Abendlandes.* C.H. Beck, Munich.

Flashar, H. (2016). *Hippokrates: Meister der Heilkunst.* C.H. Beck, Munich.

Frede, D. (2021). Platon. In *Aristoteles-Handbuch*, Rapp et al. (eds), 16–33.

Friedrich, H.-V. (1968). *Thessalos von Tralles.* SPHS, Meisenheim.

Garofalo, I. (1988). *Erasistrati fragmenta.* Giardini, Pisa.

Garofalo, I. (1991). *Galeno. Procedimenti anatomici.* Rizzoli, Milan.

Garofalo, I. (1993). Figure della medicina ellenistica. In *Lo spazio letterario della Grecia antica. Volume I: La produzione e la circolazione del testo, volume II: L'Ellenismo*, Cambiano, G., Canfora, L., Lanza, D. (eds). Salerno Editrice, Rome, 345–368.

Garofalo, I. (2001). Note filologiche sull'anatomia di Galeno. *Aufstieg und Niedergang der römischen Welt*, 37(2), 1790–1833.

Garofalo, I. and Vegetti, M. (1978). *Galeno. Opere scelte.* Unione Tipografico-Editrice Torinese, Turin.

Haase, W. (1994). *Aufstieg und Niedergang der römischen Welt. Geschichte und Kultur Roms im Spiegel der neueren Forschung. Teil II: Principat. Band 37: Philosophie, Wissenschaften, Technik. - 2. Teilband: Wissenschaften (Medizin und Biologie [Forts.]).* De Gruyter, Berlin/New York.

Höffe, O. (2005). *Aristoteles-Lexikon.* Kröner Verlag, Stuttgart.

Holmes, B. (2018). Body. In *The Cambridge Companion to Hippocrates*, Pormann, P.E. (ed.). Cambridge University Press, Cambridge, 63–88.

Joly, R. (1977). *Corpus Hippocraticum. Actes du colloque hippocratique de Mons (22–26 septembre 1975).* Université de Mons, Mons.

Jones, P.M. (1997). *Materia medica: miniature medievali.* British Library/Centro Tibaldi, Noventa Padovana.

Jouanna, J. (1992). *Hippocrate.* Fayard, Paris.

Jouanna, J. (2021). Hippocrate. In *Le savoir grec*, Brunschwig, J. et al. (eds), 772–785.

Jouanna, J., Anastassiou, A., Guardasole, A. (2016). *Hippocrate. Tome IV, 1re partie: Épidémies I et III.* Les Belles Lettres, Paris.

Kudlien, F. (1979a). Praxagoras. In *Der Kleine Pauly: Lexikon der Antike.* Band 4, Deutscher Taschenbuch Verlag, Munich, 1122.

Kudlien, F. (1979b). Diokles von Karystos. In *Der Kleine Pauly: Lexikon der Antike.* Band 2, Deutscher Taschenbuch Verlag, Munich, 52–53.

Kudlien, F. (1979c). Erasistratos. In *Der Kleine Pauly: Lexikon der Antike*. Band 2, Deutscher Taschenbuch Verlag, Munich, 343–344.

Kudlien, F. (1979d). Herophilos 1. In *Der Kleine Pauly: Lexikon der Antike*. Band 2, Deutscher Taschenbuch Verlag, Munich, 1109–1110.

Kühn, C.G. (1825). *Claudii Galeni opera omnia,* 19 volumes. Cnobloch, Leipzig.

Kullmann, W. (2021). Biologie, Zoologie. In *Aristoteles-Handbuch*, Rapp et al. (eds), 106–115.

Lanza, D. and Vegetti, M. (2018). *Aristotele. La vita. Ricerche sugli animali, Le parti degli animali, La locomozione degli animali, La riproduzione degli animali, Brevi opere di psicologia e fisiologia (Parva naturalia), Il moto degli animali*. Bompiani, Milan.

Lennox, J.G. (2002). *Aristotle, On the Parts of Animals, I-IV*. Oxford University Press, Oxford.

Littré, E. (1839–1861). *Œuvres complètes d'Hippocrate*, 10 volumes. J.-B. Baillière, Paris.

Lloyd, G.E.R. (1996). Observation et recherche. In *Le savoir grec*, Brunschwig, J. et al. (eds), 260–289.

Manetti, D. (2011). *Anonymus Londiniensis. De medicina*. Verlag C.H. Beck, Berlin/New York.

Manetti, D. and Roselli, A. (1994). Galeno commentatore di Ippocrate. *Aufstieg und Niedergang der römischen Welt*, 37(2), 1529–1635, 2071–2080.

Marchetti, F. (2010). Le illustrazioni dei testi *Sulle articolazioni* (Περὶ ἄρθρων πραγματεία) di Apollonio di Cizio e *Sulle fasciature* (Περὶ ἐπιδέσμων) di Sorano di Efeso. In *La collezione di testi chirurgici di Niceta: Firenze, Biblioteca Medicea Laurenziana, Plut. 74.7 – Tradizione medica classica a Bisanzio*, Bernabò, M. (ed.). Edizioni di Storia e Letteratura, Rome, 55–90.

Mewaldt, J. (1914). *Galeni in Hippocratis de natura hominis commentaria III*. Corpus Medicorum Graecorum V 9,1, Teubner, Berlin/Leipzig.

Nickel, D. (2009). Pseudepigraphisches zur Anatomie bei Rufus von Ephesos. In *Antike Medizin*. De Gruyter, Berlin/New York, 63–74.

Nutton, V. (2020). *Galen: A Thinking Doctor in Imperial Rome*. Routledge, London/New York.

Oser-Grote, C. (2021a). Anatomê/Anatomie. In *Aristoteles-Lexikon*, Höffe (ed.), 43–44.

Oser-Grote, C. (2021b). Pneuma/Pneuma, Atem. In *Aristoteles-Lexikon*, Höffe (ed.), 467–468.

Pellegrin, P. (1996a). Médecine. In *Le savoir grec*, Brunschwig, J. et al. (eds), 470–492.

Pellegrin, P. (1996b). Aristote. In *Le savoir grec*, Brunschwig, J. et al. (eds), 628–655.

Perilli, L. (2017). Il bestiario di Galeno. In *La zoologia di Aristotele e la sua ricezione dall'età ellenistica e romana alle culture medievali. Atti della X "Settimana di Formazione" del Centro GrAL, Pise, 18–20 novembre 2015*, Sassi, M.M. et al. (eds). Pisa University Press, Pisa, 107–143.

Perilli, L., Brockmann, C., Fischer, K.-D., Roselli, A. (eds) (2011). *Officina Hippocratica. Beiträge zu Ehren von Anargyros Anastassiou und Dieter Irmer*. De Gruyter, Berlin/ Boston.

Rapp, C.H.R. and Corcilius, K. (eds) (2011, new edition 2021). *Aristoteles-Handbuch: Leben, Werk, Wirkung*. Verlag Metzler, Stuttgart/Weimar.

Rocca, J. (2008). Anatomy. In *The Cambridge Companion to Galen*, Hankinson, R.J. (ed.). Cambridge University Press, Cambridge, 242–262.

Roselli, A. (2011). I maestri di Galeno, Galeno come maestro. In *L'insegnamento delle technai nelle culture antiche. Atti del convegno. Ercolano, 23–24 marzo 2009*, Roselli, A. and Velardi, R. (eds). Fabrizio Serra Editore, Pisa/Rome, 15, 53–70.

Russo, L. (2011). *La rivoluzione dimenticata. Il pensiero scientifico greco e la scienza moderna*. Feltrinelli, Milan.

Sideras, A. (2011). Oreibasios und Pseudo-Rufus. Ihr gemeinsames Exzerpt aus der Schrift des Rufus von Ephesos *Über die Benennung der Körperteile des Menschen* und ihr Abhängigkeitsverhältnis. In *Officina Hippocratica*, Perilli et al. (eds), 221–254.

von Staden, H. (1989). *Herophilus. The Art of Medicine in Early Alexandria*. Édition traduction et essais, Cambridge.

Steckerl, F. (1958). *The Fragments of Praxagoras of Cos and his School*. Brill, Leiden.

Steger, F. (2021). *Galenos. Arzt und Philosoph: fünf autobiographische Schriften*. Hiersemann, Stuttgart.

9

Greek Medicine: Therapeutics

Laurence TOTELIN
Cardiff University, UK

9.1. Introduction

> During the same times the Art of Medicine was divided into three parts: one being that which cures through diet, another through medicaments, and the third by hand. The Greeks termed the first Διαιτητική (*diaitētikē*, dietetics), the second Φαρμακευτική (*pharmakeutikē*, pharmacology), the third Χειρουργία (*cheirourgia*, surgery) (Celsus, *On Medicine*, prooemium 12).

As the Roman encyclopedist Celsus (fl. c. 15–35 CE) succinctly expressed in the above quotation, there were three branches of medicine in antiquity: dietetics, pharmacology and surgery. Celsus, then, equated the art of medicine with therapeutics, a term derived from the Greek word *therapeia*, which encompassed both the notions of "care" and "cure". Ancient therapeutics dealt both with the healthy and the sick body and indeed also with the mind (van der Eijk 2008). Dietetics, which developed later than the other branches of therapeutics, aimed principally at maintaining good health and was perceived as good balance within the body. Pharmacological treatments, for their part, were considered preferable to surgical ones, as stated in the final Hippocratic aphorism: "Those diseases that medicine does not cure are cured by the knife. Those that the knife does not cure are cured by fire. Those that fire does not cure must be considered incurable" (*Aphorisms* 7.87, see Jouanna 1999, pp. 154–176). Dietetic and pharmacological treatments could, to an extent, be carried out by laypeople or by laypeople supervised by physicians; surgery, on the other hand, had to be performed by

History of Life Sciences and Medicine 1,
coordinated by Lorenzo PERILLI.

specialists. For that reason, the physician in the ancient imagination was the person who "cuts and burns" (*temnein kai kaiein* in Greek), recognizable in visual arts by the presence of the cupping vessel and the scalpel used in venesection (Figures 9.1 and 9.2) (Jouanna 1999; Baker 2016). In addition to signaling expertise in therapeutics, these tools could also serve as status symbols: while the satirist Lucian of Samosata (2nd century CE) was exaggerating for literary effects, some people did possess "ivory pill boxes, silver cupping glasses and gold-inlaid scalpels" (*Against the Ignorant Book Collector* 29). This chapter offers an overview of the three ancient branches of therapeutics, drawing mainly on medical texts, composed between the 5th and 4th centuries BCE (the Hippocratic Corpus, a heterogeneous collection of some sixty texts; see Craik 2015) and the 7th century CE (Paul of Aegina), as well as archaeological material such as remains of pharmacological preparations and surgical instruments.

Figure 9.1. *A sat physician holds a scalpel in his hands to treat a standing patient. Between them is a cupping vessel. Athenian red-figure aryballos attributed to the Clinic Painter, c. 480–470 BCE, currently in Le Louvre, Paris, CA2183. ©2011 RMN-Grand Palais (musée du Louvre)/Stéphane Maréchalle. For a color version of this figure, see www.iste.co.uk/perilli/history1.zip*

Figure 9.2 *A sat physician examined the belly of his standing patient; on their right is a very large cupping vessel. Marble tombstone of Athenian physician Jason and his relatives, 2nd century CE, currently in the British Museum, London, 1865,0103.3. ©Carole Reddato via Wikimedia, CC BY 2.0. For a color version of this figure, see www.iste.co.uk/perilli/history1.zip*

9.2. Dietetics

The aims of ancient dietetics were to preserve health (i.e. to act prophylactically), treat mild illnesses and help restore health after more invasive pharmacological or surgical treatments.

The Greeks considered dietetics to be a relatively late invention, dating to the 6th or 5th century BCE, and attributable to one of several authorities: the philosopher Pythagoras of Samos (6th century BCE); the athletes' trainer Herodicus of Selymbria (5th century BCE) or the famous physician Hippocrates of Cos (5th–4th centuries BCE) (Longrigg 1998, pp. 146–149). Modern scholars have placed much emphasis on the innovative character of Greek dietetics, seeing it as the most prestigious branch of ancient therapeutics (overview in Totelin 2018); however, it is important to note that dietetics was not without its detractors in antiquity – both Plato (*Republic* 406d–e) and Aristophanes (*Frogs* 939–943) criticized it.

Dietetics is well represented in the Hippocratic Corpus: texts such as *On Regimen in Acute Diseases* and *On Regimen* focus on that branch of therapeutics, but many others mention dietetic treatments. The Hippocratic Oath stated that treatments should be dietetic in nature (*diatēmasi pasi*). This emphasis on dietetics in the writings of medical authors continued throughout antiquity. Thus, Galen (129–c. 216 CE) devoted several treatises of varying length to the topic (e.g. *On the Preservation of Health*, *On the Properties of Foodstuffs* and *On the Exercise with the Small Ball*), and two Byzantine medical encyclopedias opened with books on dietetics: Oribasius's *Medical Collections* (books 1–6, 4th century CE) and Paul of Aegina's *Medical Compendium* (book 1).

Diet, that is, food and drink intake, naturally played a key role in ancient dietetics (Wilkins 2021). Foods were thought to possess specific elementary qualities (such as being hot, cold, dry and wet), which could help restore balance in the body, on the basis that opposites cure opposites (allopathy). For example, drying foods such as fish could counteract excessive wetness in the body. Medical catalogs of foods, such as those preserved in the Hippocratic *On Regimen II* or in Galen's *On the Properties of Foodstuffs*, are important sources for the study of ancient diets: for example, they show the central place of cereals in those diets; they inform us about ancient food "taboos" (eating dog's meat was no issue for the author of *On Regimen* 2.46, but Galen is more circumspect, *On the Properties of Foodstuffs* 3.1); and they tell us about famine foods (for example, bitter vetch, see Galen, *On the Properties of Foodstuffs* 1.29). An important question for ancient dietetics was that of the difference between a food and a drug. Galen's definition is particularly neat: "Now those substances which are assimilated [by the body] are called foods; all others are called drugs" (*On Mixtures* 3.2, translation by Singer

1997). A food, then, was assimilated by the body and helped it function in its daily activities; a drug, on the other hand, could not easily be assimilated and could affect the body.

In addition to food and drink, ancient dietetics regulated other aspects of a patient's life. Exercise was to be used with caution: gentle workouts, such as walking, exercising with a small ball, or being carried in a sedan, were beneficial; vigorous exercise could be harmful, and Herodicus (probably Herodicus of Selymbria) was criticized by a Hippocratic author for killing patients suffering from fevers with his regimen of "running, much wrestling, and hot baths" (*Epidemics* 6.3.18). Dietetics also sought to control sexual intercourse, sleep, bathing and massages, and the quality of the environment in which the patient lived.

It was essential to tailor a regimen to a person's individual circumstances, gender, age and general constitution (*krasis* in Greek). Ancient authors therefore wrote about the regimens of various categories of people, such as infants and their nurses (since much could be passed to the infant through the milk), pregnant women, athletes and elderly people. Dietetic treatments, which at times went against these tailoring principles, were also recommended in the context of temple healing. They are recorded in votive inscriptions, particularly those preserved at Epidaurus, and in the orator Aelius Aristides's (117–181 CE) "autobiographical" account of illness and treatment by the god Asclepius, the *Sacred Tales*. Thus, Asclepius enjoined both a certain Cleimenes of Argos (c. 350 BCE, *IG* IV^2.121, 37) and Aelius Aristides to bathe outside in the middle of winter (*Sacred Tales* 2.19–23), a practice at which some physicians might have baulked.

9.3. Pharmacology

The Hellenistic physician Herophilus (fl. c. 280–260 BCE) asserted that *pharmaka* drugs were the "hands of the gods" (fragment 248, von Staden 1989), presumably alluding to their powerful effects. The Greek word *pharmakon* could refer both to healing drugs and to poisons, and the ancients recognized that different dosages could lead to different outcomes, ranging from alleviation of symptoms to death with drugs such as opium. The danger posed by some drugs is referred to in the Hippocratic Oath, which included clauses about abortive and lethal drugs, the interpretation of which remains contested (Leven 2018, pp. 169–179).

The ancients used many natural substances, either on their own (as "simples") or in compound remedies, in a pharmacological manner. Most of these substances were herbal, but animal and mineral substances were also featured. Many substances were easily available in most of the Mediterranean, but some had to be imported from regions further away, particularly from Arabian (for example, frankincense and myrrh) and Indian (for example, pepper) peninsulas.

No work specifically devoted to pharmacology is preserved in the Hippocratic Corpus. There are, however, allusions to the existence of such treatises, called *Pharmakitides*, in the Hippocratic treatise *On Affections*. In addition, numerous pharmacological recipes and references to the pharmacological uses of substances are preserved in the nosological and gynecological treatises of the Corpus. It is possible that some of the recipes included in the Hippocratic gynecological treatises reflect the knowledge of midwives and other healing women, appropriated by male physicians; however, it is important to stress that writing might have had a deeply transformative effect on that knowledge (King 1998, pp. 132–156; Totelin 2009, pp. 111–139). More generally, the link between folk knowledge of drugs, which was undoubtedly extensive in antiquity, and pharmacological writing is complex, as is that between oral and written transmission of knowledge (see essays in Harris 2016).

Many of the pharmacological remedies described in the Hippocratic Corpus, and beyond, are purgatives; they purge excess bodily fluids, such as phlegm, bile (black or yellow) or blood. The two main modalities of purgation were vomiting, to purge "the upper cavity of the body," and excreting, to purge "the lower cavity of the body," but purging through the vagina or the nose also occurred. The word used in Greek for those purges was *katharsis*, which had strong religious connotations. Purging was recommended for many illnesses, some of which today would be classified as mental illnesses; thus, two letters allegedly by Hippocrates deal with the administration of hellebore to treat the philosopher Democritus, who could not stop laughing (*Letters* 20–21).

The Hellenistic period (the period after the conquests of Alexander at the end of the 4th century BCE) witnessed changes in pharmacology: the number of imported substances increased, as did the number of ingredients in compound remedies, and it became more common to attribute a remedy to an authority, such as a king or a famous physician. When and how exactly those changes took place, however, is difficult to determine, as most pharmacological works written in the Hellenistic period are known only through fragments preserved in later works, especially those of Galen (Fabricius 1972); the main exceptions are the long poems on toxicology by Nicander of Colophon (2nd century BCE), *Theriaka* and *Alexipharmaka*.

Preserved pharmacological works from the 1st century CE onward belonged to several different genres (Schulze 2003). Some works focused on simple single ingredients; the most prominent in this genre was Dioscorides' *On Materia Medica*, where the drugs were organized according to a system of affinity (Riddle 1985).

Ancient readers, however, seemed to prefer to see their drugs organized alphabetically, a system adopted by Galen (in *On the Powers of Simple Drugs*) and his successors, and applied to Dioscorides' work itself, which was eventually alphabetized. While Dioscorides' work appears not to have originally been illustrated, it was later, as beautifully exemplified by the so-called "Vienna Dioscorides codex", dating to the early 6th century CE. Other pharmacological treatises focused on compound remedies. These texts could be organized either in the head-to-toe order (as Galen's *On the Composition of Drugs according to Places*) or by type of remedy (as Galen's *On the Composition of Drugs according to Kind*); some treatises focused on specific types of drugs, such as antidotes (as Galen's *On Antidotes*) or drugs that were easily procured, the *euporista* genre (as the three pseudo-Galenic treatises *On Drugs Easily Procured*) (Vogt 2009). Numerous pharmacological recipes are also preserved on papyrus; the so-called "Michigan Medical Codex" (P.Mich. 758, 4th century CE) is particularly noteworthy, as it includes annotations by the owner, thus providing information on the ways in which ancient medical compilations were read and used (Youtie 1996). Galen referred to collections of recipes in parchment codices, an early use of the codex book format (Nicholls 2010).

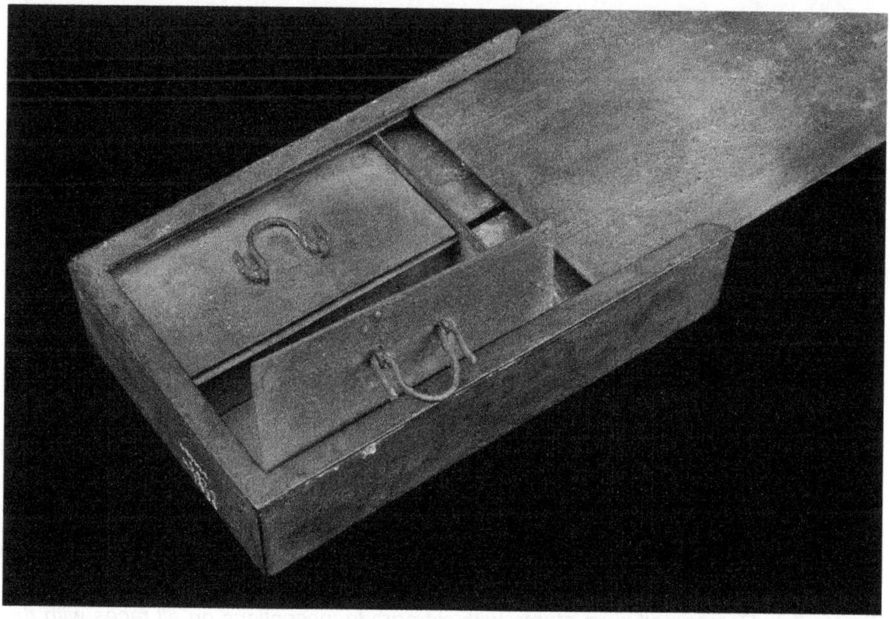

Figure 9.3. *Copy of a bronze Roman medicine box with a sliding lid and divided into compartments. Currently in the Science Museum, London, A155225. ©Wellcome Images, CC BY 4.0. For a color version of this figure, see www.iste.co.uk/perilli/history1.zip*

Ancient pharmacological treatises reveal large areas of overlap between medicine and other arts, such as cooking, cosmetics and perfumery, dyeing and painting, gem making and magic. Recipes for amulets are relatively common in ancient pharmacological texts; Soranus (1st–2nd centuries CE) noted that, while he did not believe in their efficacy, he recommended not to forbid using them "for even if the amulet has no direct effect, still through hope it will possibly make the patient more cheerful" (*Gynaecology* 2.42, translation Temkin 1956).

Archaeological findings complement our textual knowledge of ancient pharmacology. They include medicine boxes, rectangular boxes made of bronze, wood or ivory, which are divided into compartments; pharmacological containers, such as microcontainers for *lykion*, an expensive preparation that is particularly useful for the eyes; collyrium stamps, stamps used to inscribe medicinal pastes (*collyria*) recommended especially for the treatment of eye diseases; and remains of *collyria* and other medicinal substances (Gourevitch 2019, pp. 137–158) (Figures 9.3–9.4).

Figure 9.4. *Roman collyrium stamp with retrograde inscriptions on all faces with the name "Eugenius" and other information about remedies. The stamp was excavated from the bed of the Moselle River in Trier, Germany. Currently in the Science Museum, London, A629428. ©Wellcome Images, CC BY 4.0. For a color version of this figure, see www.iste.co.uk/perilli/history1.zip*

9.4. Surgery

References to surgery are found in one of the earliest Greek texts, Homer's *Iliad*, where numerous wound treatments are described (Salazar 2000). Many ancient surgeries dealt with the negative effects of warfare or sporting activities.

The short Hippocratic treatise *In the Surgery* describes the requisite elements to practice surgery: the operator, their assistants, their equipment (including bandages, to which a long section is devoted) and good lighting, whether natural or artificial. Another treatise preserved in the Hippocratic collection, *On the Physician*, covers similar grounds but is probably much later in date.

Orthopedics is well represented in the Hippocratic Corpus, with the much-admired treatise *On Joints*, which includes descriptions of various implements, which rely on technologies such as the windlass, pulley, crank and lever. One such implement would later become known as the "Hippocratic bench" (used in reductions, especially of the hips, *On Joints* 78). Other forms of surgery, however, are also present in the Hippocratic Corpus: for example, *On Head Wounds* focuses on trephination; *On Superfetation* explains how to carry out an embryotomy; *On Fistulas* and *On Hemorrhoids* describe some surgical procedures to address the common issues of fistulas and hemorrhoids, including those in women; *On Sight* depicts the cutting and cauterization of the eyelids (Witt 2018).

Like pharmacology, surgery evolved during the Hellenistic period owing to anatomical investigations carried out in the intellectual centers of the time, primarily Alexandria. However, it is difficult to determine exactly how and when those changes occurred because works written in this period are lost and known only through citations by later authors. The exception is Apollonius of Citium's *On Joints* (1st century BCE), which included illustrations, which might have served as models for the stunning illustrations in the 10th-century so-called "Nicetas codex". The encyclopedist Celsus's books 6 and 7 of *On Medicine* (1st century CE) are good sources of information on the state of surgery at the beginning of the Common Era. Celsus stressed that a surgeon needed a steady hand and an ability not to be moved by the cries of his patient (*On Medicine* 7, prooemium 4). While ancients knew the anesthetic properties of plants such as mandrake (see Dioscorides 4.75), operations were performed on patients who had not been fully anesthetized.

Galen did not write a treatise specifically devoted to surgery, but he displayed his surgical knowledge in many of his works and recounted many autobiographical anecdotes about his surgical skills (Nutton 2020, pp. 110–112). Galen did, however, write three works on venesection (*On Bloodletting against Erasistratus*; *On Bloodletting against the Erasistrateans*; *On Treatment by Bloodletting*), a practice he strongly supported to regulate natural heat in the body.

Among the Byzantine medical authors, Paul of Aegina deserves special mention for his surgery (book 6 of his *Medical Compendium*). Best known for its description of military surgery, we can also note its inclusion of operations on the sexual organs or to address secondary sexual characteristics, for example, operations to remove excess breast tissue in men (6.46), to castrate men to create eunuchs (6.78), to treat "hermaphroditism" (surgery for intersex conditions) or to remove the clitoris (6.70).

Figure 9.5. *Physician sat on a chair reading a book scroll. On his right is a cupboard on top of which there is a case of surgical instruments. Sarcophagus of a Greek physician, made in Ostia, beginning of the 4th century CE. Currently in the Metropolitan Museum of Art, New York, 47.76.1. ©The Metropolitan Museum of Art, Open Access. For a color version of this figure, see www.iste.co.uk/perilli/history1.zip*

Ancient surgical instruments are relatively common archaeological finds. Physicians sometimes chose to be interred with their surgical tools or have them represented on their funerary stele or sarcophagus (Figure 9.5). Large amounts of surgical tools have also been found in Pompeii, Rimini and Roman military forts. Scalpels (or rather their handles because the iron blades do not survive), forceps, cupping vessels and probes are very common, but there have also been findings of specula, trephines, catheters, saws, hooks, etc. (Bliquez 2015) (Figure 9.6).

Some of these tools are very ornate, which might have made them difficult to clean. Galen described the way in which he created new surgical tools: he made wax models, which he then gave to bronze smiths (*Avoiding Distress* 5).

Figure 9.6. *Photo by Giorgio Sommer (1834–1914) of Roman surgical instruments found at Pompeii and kept in the Museo di Napoli. ©Giorgio Sommer, Public Domain*

9.5. References

Baker, P.A. (2016). Images of doctors and their implements: A visual dialogue between the patient and the doctor. In *Homo Patiens: Approaches to the Patient in the Ancient World*, Petridou, G. and Thumiger, C. (eds). Brill, Leiden, 365–389.

Bliquez, L.J. (2015). *The Tools of Asclepius: Surgical Instruments in Greek and Roman Times*. Brill, Leiden.

Craik, E.M. (2015). *The 'Hippocratic' Corpus: Content and Context*. Routledge, London.

van der Eijk, P.J. (2008). Therapeutics. In *The Cambridge Companion to Galen*, Hankinson, R.J. (ed.). Cambridge University Press, Cambridge, 283–303.

Fabricius, C. (1972). *Galens Exzerpte aus älteren Pharmakologen*. De Gruyter, Berlin.

Gourevitch, D. (2019). *Pour une archéologie de la médecine romaine*. De Boccard, Paris.

Harris, W.V. (2016). *Popular Medicine in Graeco-Roman Antiquity: Explorations*. Brill, Leiden.

Jouanna, J. (1999). *Hippocrates. Translated from the French by M.B. DeBevoise*. The Johns Hopkins University Press, Baltimore.

Keyser, P.T. and Irby-Massie, G L. (2008). *The Encyclopedia of Ancient Natural Scientists: The Greek Tradition and Its Many Heirs*. Routledge, London.

King, H. (1998). *Hippocrates' Woman: Reading the Female Body in Ancient Greece*. Routledge, London.

Leven, K.-H. (2018). Ethics and deontology. In *The Cambridge Companion to Hippocrates*, Pormann, P.E. (ed.). Cambridge University Press, Cambridge, 152–179.

Longrigg, J. (1998). *Greek Medicine from the Heroic to the Hellenistic Age: A Source Book*. Duckworth, London.

Nicholls, M. (2010). Parchment codices in a new text of Galen. *Greece & Rome*, 57, 378–386.

Nutton, V. (2020). *Galen: A Thinking Doctor in Imperial Rome*. Routledge, London.

Riddle, J.M. (1985). *Dioscorides on Pharmacy and Medicine*. University of Texas Press, Austin.

Salazar, C.F. (2000). *War Wounds in Graeco-Roman Antiquity*. Brill, Leiden.

Schulze, C. (2003). *Die pharmazeutische Fachliteratur in der Antike: eine Einführung*. Göttingen, Duehrkohp.

Singer, P. (1997). *Galen: Selected Works*. Oxford University Press, Oxford.

von Staden, H. (1989). *Herophilus: The Art of Medicine in Early Alexandria*. Cambridge University Press, Cambridge.

Temkin, O. (1956). *Soranus; Gynecology. Translated with an Introduction*. The Johns Hopkins Press, Baltimore.

Totelin, L.M.V. (2009). *Hippocratic Recipes: Oral and Written Transmission of Pharmacological Knowledge in Fifth-and Fourth-Century Greece*. Brill, Leiden.

Totelin, L.M.V. (2018). Therapeutics. In *The Cambridge Companion to Hippocrates*, Pormann, P.E. (ed.). Cambridge University Press, Cambridge, 200–216.

Vogt, S. (2009). Drugs and pharmacology. In *The Cambridge Companion to Galen*, Hankinson, R.J. (ed.). Cambridge University Press, Cambridge, 304–322.

Wilkins, J. (2021). Food. In *A Cultural History of Medicine in Antiquity*, Totelin, L.M.V. (ed.). Bloomsbury, London, 44–67.

Witt, M. (2018). Surgery. In *The Cambridge Companion to Hippocrates*, Pormann, P.E. (ed.). Cambridge University Press, Cambridge, 217–245.

Youtie, L.C. (1996). *The Michigan Medical Codex*. Scholars Press, Atlanta.

10

Zoology in the Ancient Greek World and Its Reception in the Islamic and Latin Worlds

Tommaso ALPINA[1,2] and Andrea FALCON[3,4]
[1] *Università di Pavia, Italy*
[2] *American University of Beirut, Lebanon*
[3] *Concordia University, Montreal, Canada*
[4] *Università degli Studi di Milano, Milan, Italy*

10.1. Zoology before Plato and Aristotle

Before Aristotle, there was no single work, let alone set of works, dealing specifically with the study of animals. This does not mean that there was no scientific interest in the study of animals. On the contrary, there is clear evidence that early Greek thinkers were actively engaged in the study of animals. This study was offered in the context of a narrative account of the natural phenomena ab initio. The beginning of the narration coincided with the beginning of the world, and the overall account took the form of a narration of how the present world order (*cosmos*) emerged. In short, it took the form of *cosmogony*.

Zoology in the ancient Greek world (sections 10.1 to 10.3) and the coda on Dante were written by Andrea FALCON. Tommaso ALPINA is responsible for the Islamic and Latin receptions of ancient Greek zoology (sections 10.4 to 10.6).

History of Life Sciences and Medicine 1,
coordinated by Lorenzo PERILLI.

This cosmogonic framework imposed serious constraints on how the study of animals was conducted. First, this study was part of an account of how all living beings, including animals, came into existence. In addition, it was given in the style of a story of the emergence of living beings from non-living matter. In short, it took the form of a *zoogony* (from the Greek *zôia*, which here means “living beings” rather than “animals”). Ancient zoogonies were committed to what came to be known as *spontaneous generation*, namely, the spontaneous emergence of life from earth and water. This style of explanation had the advantage of relating to the ancient myths of the autochthonous birth of cultural heroes (Campbell 2014).

Empedocles’ account of the origin of living beings is arguably the best-known example of this style of explanation. We still have a few fragments from his lost poem on nature. They have come down to us because they captured the imagination of ancient readers:

> (A) “From it <*sc.* the earth> blossomed many faces without necks, and naked arms wondered about, bereft of shoulders, and eyes roamed about alone, deprived of brows” (Simplicius, *Commentary on Aristotle’s On the Heavens* 586.12-587.1-2 = Diels/Kranz 31B57, trans. Laks/Most).
>
> (B) “The limbs, still solitary-membered […], were wondering desiring to mix with one another” (Simplicius, *Commentary on Aristotle’s On the Heavens* 587.18-19 = Diels/Kranz 31B58; trans. Laks/Most).
>
> (C) “Many grew double of faces and double of chests, races of man-faced cattle, while others sprang up inversely, creatures of cattle-headed men, mixed here from men, there creatures of women fitted with shadowy genitals” (Aelian, *The Nature of Animals* 16.29 = Diels/Kranz 31B61, trans. Laks/Most, slightly modified).

Empedocles explained the formation of living bodies as the outcome of a zoogonic process that unfolds in four steps (Aëtius, *Placita* V 19.6 = Diels/Kranz 31A72). While the first generation resulted in the formation of proto-organs enjoying the existence of their own (the neckless faces, naked arms bereft of shoulders, eyes roaming alone in need of foreheads mentioned in A and B), the second and third generations gave rise to whole organisms out of these proto-organs[1]. These organisms could not survive because they were randomly assembled (most famously, the cattle-headed men or man-faced cattle in C). Even when they managed to survive, these living beings could not reproduce themselves

1 We speak of “proto-organs” because the parts envisioned in the first generation are not yet part of the whole organism.

sexually. It is only with the fourth generation that the whole organisms could not only survive but also reproduce sexually. The details of Empedoclean story are open to more than one reading[2]. What matters is that Empedocles regarded life as an emergent phenomenon and explained both human and non-human animals as products of random processes rather than as the outcome of an intelligent plan.

At first sight, this fascinating account of species formation may sound like a Darwinian theory of evolution through natural selection. Upon reflection, however, the differences are more significant than the similarities. Empedocles did not have a natural mechanism that could account for the formation and propagation of animal species; he only envisioned a negative role for natural selection to the extent that he stipulated that any spontaneously formed whole unfit to survive did not survive (Henry 2016).

The Hippocratic work *On Fleshes* (*De carnibus*) cast further light on how the emergence of life was explained before Aristotle. The unknown author of this work was presumably a physician eager to use the most exciting results achieved in the ancient Greek study of nature. At the outset of his discourse, he makes an explicit effort to establish contact with the best scientific theory of his time:

> (D) In this discourse (*logos*), I will employ assumptions that are generally held—both those of my predecessors and my own—since it is necessary to establish a shared starting point if I wish to compose this discourse about the art of medicine. About what is in the heavens I have no need to speak, except insofar as is necessary in order to explain *how the human being and the other animals are formed and come into being*, what life (*psychê*) is, what health and sickness are, what is evil and what is good in the human being, and where its death comes from. From here on, then, I present views that are my own" ((Hippocrates), *On Fleshes* 1, trans. P. Potter, with minor changes).

Our Hippocratic author could not be more forthcoming on his argumentative strategy and goals. His starting point is a widely shared one; for this reason, it is also the most appropriate starting point for a medical discourse that has the stated ambition of explaining the formation of the various parts of the human and non-human body. This starting point is the emergence of the present world order as the result of a process of separation of what was originally mixed. The focus of this cosmogonic account is on the emergence of various parts of the human body, but reference to animals suggests that the results obtained in the study of the human body are, in principle, transferable to non-human living beings.

2 The most recent attempt at a reconstruction of the zoogonic process is Wellmann (2020, pp. 128–158).

The separation mechanism of what was originally mixed is rotation caused by an original vortex. Invoking a prescosmic vortex to explain the separation of the various parts of the *cosmos* was a relatively common device in the context of the early Greek study of nature. The formation of each of the relevant bodily organs is in turn described as the result of the action of the trapped heat on the surrounding substances. Depending on the substances this heat came into contact with, different parts of the body were formed. These parts were spontaneously formed due to the action of heat on the surrounding substances.

The order in which the various organs were reportedly formed is far from clear. However, even if we adopt a charitable approach to the text, we still do not see how a living body (either a human or a non-human body) was subsequently formed. Our author is in fact expected to explain how bones, cords and vessels – and indeed flesh as well as the other bodily parts – come together in a human or non-human body, since the explanation of the formation of the various parts is not by itself an explanation.

Let us take stock: like Empedocles, the Hippocratic author of the treatise *On Fleshes* took life to be an emergent phenomenon. Like Empedocles, he explained the formation of various organs before the whole organism was explained. Unlike Empedocles, he did not explain how the various parts came together to form a living body. His silence is not a coincidence; rather, it is an implicit admission that his explanatory strategy runs into serious limitations.

10.2. Plato

When we reflect on the conceptual difficulties that the earliest Greek accounts of spontaneous generation faced and could not overcome, we can only be sympathetic to the reform advocated by Plato. In *Timaeus*, Plato adopts a cosmogonic framework. Like his predecessors, he offers an account of the emergence of the present world order from the beginning. Unlike his predecessors, Plato thinks that no adequate explanation of the present world order is possible unless intelligence (*nous*) is injected into the cosmogonic account. By arguing that the world order is a product of intelligence, Plato adopts Anaxagoras's *nous* (intelligence). However, he goes beyond anything Anaxagoras said to the extent that he conceives of intelligence in terms of practical rationality bound by the rules of craftsmanship. In short, Plato not only adopts the Anaxagorean *nous* but also transforms it into something new: a divine craftsman who is not beyond the rules of productive reason but rather applies them to create our *cosmos*.

Timaeus is centrally concerned with the creation of the present world order down to the creation of human beings. *Tim.* 27 A combined with *Tim.* 90 E suggest that the short and highly selective account of the creation of non-human animals is a coda to the main cosmogonic story. However, this does not mean that it is not possible to say something definite as to how Plato would promote a full study of animals within a cosmogonic framework. Like all previous cosmogonic accounts, Timaeus offers a *causal account* that identifies and discusses the main causal factors responsible for the emergence of the present world order. For Plato, the *primary cause* is intelligence, which is conceived in terms of productive rationality. In dealing with the human eye and the mechanism of vision, Plato makes it clear that another kind of cause is needed for an adequate explanation of our *cosmos*. He refers to it as an *auxiliary cause*. This is the cause that all his predecessors (no one excluded) discussed. Plato thinks that they were all mistaken to the extent that they focused on this cause to the exclusion of intelligence:

> (E) All of the above are among the auxiliary causes (*sunaitia*) employed in the service of the god as he does his utmost to bring to completing the character of what is most excellent. But because they make things cold or hot, compact or disperse them, and produce all sorts of similar effects, most people regard them not as auxiliary causes but as the actual causes of all things. Things like these are, however, totally incapable of possessing any reason or understanding about anything. [...] So anyone who is a lover of understanding and knowledge must of necessity pursue as primary causes those that belong to intelligent nature, and as secondary all those belonging to things that moved by others and that set still others in motion by necessity. We, too, must do likewise: we must describe both type of causes, distinguishing those that possess understanding and thus fashion what is beautiful from those which, when deserted by intelligence produce only haphazard and disorderly effects every time (Plato, *Tim.* 46 D-E, trans. D. J. Zeyl).

Here, Plato is not content to recommend the integration of intelligence and the auxiliary cause into a single causal account. By his lights, intelligence has explanatory priority over the auxiliary cause. Consider Plato's explanation of human vision. For Plato, the ultimate purpose of human vision is to observe the revolutions of day and night, the month, and the year to determine their number and time. By observing the ordered revolutions of the fixed and wondering stars, the circular motions of the rational soul, disrupted after their embodiment in human bodies, may be eventually restored[3]. In this context, the ultimate purpose of human vision is to contribute to restoring rationality via a study of the natural world (with a focus on the study of heavenly objects).

For Plato, *an intelligent plan* is implemented via the creation of the human body. This creation is the task of the lesser gods. While these gods are not identical to the divine craftsman, they imitate his creative power; they too are bound by the rules of practical rationality in executing the plan. In creating the human eye, they consider a few facts without which the plan could not be implemented. First, the mechanism of vision requires a special kind of fire that does not burn but spreads a gentle light. For another, the fabric of the eye must be fine to let this go through. The details of Plato's explanation of the creation of the eye do not matter. What matters is that divine intelligence, which is now located in the lesser gods, produces the desired result – a human eye that performs its expected task, seeing— by working together with the auxiliary causes.

The above passage ends with the claim that the causes that are collectively named necessity (*anankê*) "deserted by intelligence, produce only haphazard and disorderly effects every time". Plato does not conceive of necessity as a collection of inert materials ready to be used for the sake of implementing an intelligent plan. He thinks that necessity is an independent kind of reality with its own distinctive motions. Left on its own, this reality never produces order or structure. Rather, it produces only haphazard and disorderly effects *every time*. To understand how radical this view is, we may briefly consider what is currently known as the infinite monkey theorem. This theorem states that a monkey hitting keys at random on a typewriter keyboard for an infinite amount of time will type any given finite text an infinite number of times. Plato denies that such an outcome is possible. Assuming that our monkey is not a purposive causal agent but is only the personification of necessity, there is no way in the world (at least for Plato) that our monkey can produce meaningful text out of random typing. Very few today would follow Plato down this road, even though many would agree that his predecessors failed to offer an adequate explanation of the emergence of life out of random processes.

10.3. Aristotle

Aristotle made zoology a cornerstone of his philosophical thought. The writings concerning the study of animals constitute more than 25% of the surviving

3 In the *Timaeus*, thinking is conceived in terms of circular motions. These motions are disrupted upon the descent of the rational soul into the body.

Aristotelian corpus. In the two-volume edition of Aristotle's works produced by Bekker in 1830–1831 – the first truly critical edition of Aristotle – the study of animals comprises 386 pages out of a total of 1,462. When we concentrate on natural philosophy, this number becomes even more staggering: 386 pages out of a total of 605 (approximately 60% of the surviving works).

Like Plato, Aristotle thinks that we cannot provide an adequate explanation of living beings in terms of necessity – namely, by invoking the hot and the cold and the other physical properties of things. Like Plato, Aristotle adopts a *teleological approach* to their study. Unlike Plato, however, Aristotle does not think that goals or plans can be imposed on nature from the outside. This amounts to an outright rejection of the account offered in *Timaeus*. For Aristotle, positing an intelligent cause of the sort envisioned by Plato amounts to positing a self-mover at the beginning of the causal process. For Aristotle, this move is unsatisfying because it ends up making self-motion a primitive fact. By his lights, the first principle must be an unmoved mover rather than a self-mover: it must be a principle that imparts motion while it is not itself in motion.

This commitment leads to a radical departure not only from *Timaeus* but also from the cosmogonic style of explanation adopted by all of his predecessors. Aristotle himself says so in an important methodological passage in which he tells us that *all his predecessors* (no one excluded) attempted to explain *X* by giving an account of how *X* has come into existence (where *X* can be the whole-world order or anything in it). For Aristotle, the sound methodological approach is the opposite:

> (F) We must also consider whether we should follow the procedure of our predecessors, by studying how a thing naturally comes to be rather than how it is. For it matters quite a bit which procedure we follow. Now, it would seem that in the case of coming to be we should begin from how things are; for, just as we said before, we must begin with the appearances (*phainomena*) about a given kind of thing and then go on to state their causes. For in the case of building also, this comes about because the form of a house is of this sort, whereas it is not true that a house is of this sort because this is how it comes to be; for coming to be is for the sake of being, and not being for the sake of coming to be (*Parts of Animals* I 1, 640a10-19; trans. T. Irwin and G. Fine).

The methodological insight offered in the passage shapes Aristotle's whole study of animals. Aristotle illustrates his approach to the study of animals by using housebuilding as an example. Just outside of our passage, he also provides a historical example when he recalls how Empedocles explained the formation of the human spinal column. Empedocles argued that the vertebrae are formed when the spinal column is twisted as the baby turns in the womb of the mother. Aristotle is not impressed by this explanation, even though he does not offer an alternative account of the formation of the human spinal column. He noted that a spinal column, with its distinctive shape and number of vertebrae, is already present in the generator. Therefore, explaining the spinal column by focusing on the process of its formation is a mistake. The explanation of the spinal column must take its lead from the functional role that a fully formed spinal column plays in a fully developed human being (the generator). This is because the spinal column is articulated into the vertebrae such that its formation takes place in the way it does and not vice versa.

The implications of this observation become fully apparent when we reflect on how the study of animal life is organized in the two main causal treatises that have come down to us. For Aristotle, an account of the complex processes resulting in the coming to be of an animal (offered in the *Generation of Animals*) is possible only after the systematic study of animals and their bodily parts (advanced in *Parts of Animals*) is in place. From the nature of what is to be produced (the animal), certain things follow (out of conditional necessity) regarding how the animal must come into existence. That is also why the parts of the animals come into being in the way and manner they do (Gotthelf 2012a, p. 91). This point can be illustrated by means of housebuilding: what the housebuilders ought to do in the course of housebuilding is intelligible only when the outcome is considered. To grasp the outcome first, the finished house must be in place to understand the process of housebuilding.

The programmatic passage quoted above contains another important insight: the scientific enterprise proceeds in stages, and the two main stages of any scientific inquiry are the collection of the relevant data (the *phainomena*) followed by their explanation. The study of animals is no exception to this rule. The Greek manuscript tradition has transmitted a treatise that is entirely devoted to the collection and presentation of zoological data – namely, the History of Animals. "History" is an imperfect translation of Greek historia, which means "research" or "inquiry". For Aristotle, the research or inquiry in question is not yet concerned with the explanation of zoological data. Rather, it is concerned with their collection and presentation in view of their subsequent explanation. Therefore, the History of Animals is best understood as representing a preexplanatory stage of inquiry (Gotthelf 2012b).

When we keep in mind how Aristotle conceives of scientific inquiry, we see that the correct order in which the three main zoological treatises are to be read is *History of Animals*, *Parts of Animals* and *Generation of Animals*. While the

History of Animals marks the pre-explanatory (*hoti*) stage of inquiry, *Parts of Animals* and *Generation of Animals* jointly contribute to the explanatory (*dioti*) stage of inquiry. This does not mean that Aristotle wrote these works in this very order. A careful study of the zoological corpus led David Balme to formulate a developmental hypothesis, according to which the *Historia Animalium* was the last of the zoological writings to be written by Aristotle (Balme 1987, p. 18). This hypothesis, known as "the Balme hypothesis" (Lennox 1996), is not universally accepted.

10.4. The transition from the Late Ancient to the Arabic reception of Aristotle's zoology

Despite the efforts Aristotle made to offer a scientific study of animals, his zoological writings remained at the margins of the philosophical reflection in the Hellenistic and post-Hellenistic periods, the only exceptions being the summary of Aristote's study of animals composed by Aristophanes of Byzantium (d. 180 BCE) and the compendium of Aristotelian philosophy written by Nicolaus of Damascus (*fl.* 1st century BCE). Alexander of Aphrodisias (which was active in the late 2nd and early 3rd centuries CE), regarded in antiquity and beyond as the commentator of Aristotle *par excellence*, did not engage with Aristotelian zoology. Likewise, the subsequent commentary tradition (4th–6th century CE) shows a *selective approach* to Aristotle's program of natural philosophy. In this approach, there was no room for the study of animals (Falcon 2021). In the classification of Aristotelian writings, at the beginning of their commentaries on the *Categories*, ancient Neoplatonic commentators placed Aristotle's zoological writings among his so-called "intermediate works" (Hadot 1990, pp. 69–70). By their lights, these writings have technical rather than philosophical contents. This explains why none of them composed a commentary on these writings. In the 12th century, Michael of Ephesus partially filled this gap by writing commentaries on *Parts of Animals* and *Generation of Animals* as well as on two Aristotelian works concerning animal motion (*On the Motion of Animals* and *On the Progression of Animals*) (Cerami and Falcon 2014, p. 45).

Although he did not write specifically on animals, Galen of Pergamon (129–ca. 216 CE) was inspired by Aristotle's zoological works, especially *Parts of Animals.* Two examples of Galen's debt to Aristotle are the use of a *teleological approach* to anatomical exposition in *On the Usefulness of the Parts of the Body* and the identification of three levels of bodily composition (elements/elemental qualities, humors and *homogeneous* parts, and organs) in *Elements According to Hippocrates* (Singer 2021).

The reception of Aristotle's study of animals in the Islamic world marked a decisive trend inversion. The translation of a significant portion of Aristotelian zoological writings attests to this change of mind. Aristotle's *History of Animals*, *Parts of Animals* and *Generation of Animals* were translated into Arabic in the first half of the 9th century. They circulated as unitary work consisting of 19 treatises (*maqālāt*) under the title *Book of Animals* (*Kitāb al-Ḥayawān*) or *Book of the Natures of Animals* (*Kitāb Ṭabā'i' al-Ḥayawān*). Treatises I–X contain *History of Animals* (with the inclusion of the tenth book, usually considered spurious), treatises XI–XIV correspond to *Parts of Animals*, and treatises XV–XIX convey the contents of *Generation of Animals*. Vague allusions aside, there is no trace of an Arabic translation of the two Aristotelian works on animal motion. In *The Catalogue* (*al-Fihrist*), the bio-bibliographer Ibn al-Nadīm ascribes the translation of the three main zoological writings to Yaḥyā ibn al-Biṭrīq (*fl. ca.* 835 CE), who was active in the circle of the philosopher al-Kindī. This authorship is currently debated. Al-Nadīm also attests to the circulation of an Arabic translation of Nicolaus of Damascus' compendium of Aristotle's zoology, which he ascribes to Abū ʿAlī ibn Zurʿa (d. 1008 CE)[4].

Al-Kindī (d. *ca.* 870 CE) and al-Fārābī (d. 950 CE), the first two representatives of the reception and reworking of Greek philosophy (*falsafa*) in the Islamic world, are acquainted with this translation: their classifications of the Aristotelian writings list the *Book of Animals* among his works on natural philosophy. However, they did not embark on a specific investigation of animals. Avicenna (Ibn Sīnā, d. 1037 CE) was the first to include zoology in a philosophical project set out to study nature.

10.5. Medicine and zoology in Avicenna: tradition and innovation

The *Book of Animals* (*Kitāb al-Ḥayawān*) is Avicenna's only work on zoology. It is the eighth, last, and longest section of the part devoted to natural philosophy in the *Book of the Cure/Healing* (*Kitāb al-Šifā'*), which is Avicenna's magnum opus. Written in approximately 1027, the *Book of Animals* marks the conclusion of the composition of the *Book of the Cure*. Its opening lines contain some information about the structure of the work as a whole and its relation to the main source (Aristotle):

> (G) Let us now talk about animals by following in all this book the *first teaching* (*al-taʿlīm al-awwal*) as a model, except in the case of the anatomy of the organs of the human being – actually, we prefer to put together anatomy (*al-tašrīḥ*) and [its] use (*al-manfaʿa*) in one

4 Further information on zoological writings available in the Arabic tradition in Kruk (2003).

> single place – and in the case of few [other] things. Then, we shall cut off in terms of information that which he (*sc.* Aristotle) was prolix about. We shall mention from the [preceding] theoretical discourse what is appropriate for our opinion and our collection of these sections (*Book of Animals* I, 1, 1.10–13).

Avicenna announced that he was going to adopt Aristotelian writings as a model. As a rule, Avicenna referred to Aristotle as the *First Teacher* (*al-muʿallim al-awwal*) and to his works as the *First Teaching*, as in this case. By using these expressions, Avicenna wanted to stress not only the chronological priority of Aristotle but also his philosophical primacy. This does not mean, however, that Avicenna followed Aristotle's teaching unoriginally. In contrast, he felt free to update Aristotle as he saw fit. In the passage above, for example, Avicenna revealed with us his plan to relocate the exposition of human anatomy and a few other unspecified topics from zoology to some other unspecified work (most likely, his *Canon of Medicine, Qānūn fī l-ṭibb*)[5], as well as his intention to streamline Aristotle's teaching when he perceived this to be too lengthy.

Although Avicenna's *Book of Animals* consists of 19 treatises, such as the Aristotelian *Book of Animals* available to him in Arabic translation, this similar arrangement is purely cosmetic. Consider, for the sake of illustration, treatises XI–XIV. In an attempt to update Aristotle's rudimentary anatomical knowledge, Avicenna replaces most of the original text of *Parts of Animals* with Galenic material transplanted from his own *Canon of Medicine* (Musallam 1987). In contrast to what Avicenna tells us in text (G), the insertion of medical material into zoology has led to a shift in focus toward the human animal. To fully appreciate this innovation, we need to keep in mind that the investigation of the organic body, including a study of its morphology and properties, was the field where medicine and natural philosophy not only intersected but also conflicted. This was also the field where anatomical and physiological progress was made after Aristotle and where it was most urgent to update Aristotle's teaching. Instances of this progress are the distinction between arterial and venous circulation and the nervous system. Both discoveries were made *after Aristotle* in Alexandria in Hellenistic times. Galen (2nd century CE) contributed greatly to disseminating these discoveries beyond antiquity. Being a physician no less than a philosopher, Avicenna acknowledged the significance of these discoveries and Galen's contribution to their dissemination (he always referred to Galen as *the Most Excellent among the Physicians, fāḍil al-aṭibbāʾ*). Avicenna also tried to overcome the perceived tensions between natural

5 In combining the study of the bodily organs with a discussion of their use, Avicenna is following his other main authority: Galen, who devoted an entire work to the study of the parts of the human body and their use (*On the Usefulness of the Parts of the Body*).

philosophy and medicine by providing a shared theoretical framework in order for the former to ground the latter. His zoology is the place where this theoretical framework is outlined. For this reason, zoology plays a central role in Avicenna's scientific thought.

In his zoological project, Avicenna integrated natural philosophy and medicine by means of two fundamental moves. First, he settled the disagreement between (natural) philosophers and physicians on common issues such as the role and function of the chief organs in the animal body (heart, brain and liver) and the female and male contributions to sexual generation (whether this contribution is the same or different, and if it is different, how the female contribution differs from the male contribution) (Alpina 2022). Second, he introduced four humors (blood, phlegm, yellow bile and black bile) as the proximate material constituents of the organic animal body (Alpina 2023). Treatises I–X and XV–XIX are concerned with the first move, whereas treatises XI–XIV are devoted to the second move.

Avicenna favored Aristotle's cardiocentrism over Galen's tripartite model that harks back to Plato's *Timaeus*. According to the Galenic model, the heart, liver, and brain are all on equal footing and are the sources of three distinct systems in the body. Cardiocentrism may no longer seem to be a live option after the discovery of the nervous system and the consequent relevance of the brain. However, Avicenna adopted it because it fit better with his theory of the soul. The soul's oneness requires a first connection with the body through a single organ, that is, the heart, from which all psychic powers flow into their appropriate bodily organ. Moreover, the formative power (*quwwa muṣawwira*), which is in the semen, through the mediation of the heart of the embryo, contributes to the formation and shaping of all other bodily parts. For Avicenna, the heart was then the ultimate source (*munbaʿaṯ*) of the psychic powers and the ultimate origin (*mabdaʾ*) of all bodily parts via this formative power. With respect to the female and male contributions to sexual generation, Avicenna settled the disagreement between Aristotle and Galen in two steps. First, by making a concession to *experience* (*taǧriba*), he acknowledged the existence of the female semen. By doing so, Avicenna sided with Galen, who explicitly criticized Aristotle for denying the existence of female semen. However, Avicenna distinguished male semen from female semen. They are two different things with two different functions (they share the same name – *minan* – only *by equivocation*): the male semen has the active generative power which infuses the form in the embryo *qua* efficient cause, whereas the female semen has a sort of plastic power, that is, the capacity for determining and preparing the matter of the embryo; in other words, the menstrual blood, which the male semen then fashions and structures.

Avicenna accomplished the second move outlined above by replacing the theoretical exposition contained in the first book of Aristotle's *Parts of Animals* with the exposition of the medical theory of the four humors (*aḫlāṭ*): what and how many they are, how they originate, and how they enter the constitution of the homogeneous parts (bones, tissues, muscles, cartilages, etc.) as well as the instrumental parts (the organs) of the body. The humoral theory was upheld by Hippocrates and known to Avicenna via Galen's writings. His transfer of the investigation of humors from medicine to zoology fulfills a twofold purpose. First, humors serve as a bridge between the remote matter of the organic body usually discussed in natural philosophy (the elements, which give rise to the primary mixtures through the interaction of their primary qualities) and the actual bodily parts, whose states of health and disease are the primary concern of medicine. Second, by placing the exposition of humoral theory in natural philosophy, Avicenna takes the investigation of humors away from the exclusive prerogatives of medicine and brings it to the level of philosophy. With their new status of philosophical principles, humors contribute to grounding medicine which, as a practical discipline, is subordinate to natural philosophy.

10.6. The Latin tradition to Dante

Aristotle's zoological corpus made its way into the West through the translation produced by Michael Scot (ca. 1170–1236), who was active in Toledo and, from 1220, in Italy, first at the papal Curia and then at the court of Frederick II. Michael translated from Arabic into Latin, Aristotle's zoological writings as well as Avicenna's *Book of Animals*. Dedicated to Frederick II (1230), the latter is also known as *Abbreviatio Avicennae* because it consists of an abridged version and adaptation of the original rather than a faithful translation (van Oppenraay 2017, p. 405). The subsequent translations made by William of Moerbeke directly from the Greek did not supersede those produced by Michael, which continued to be used well into the 14th (and even the 15th) century.

In the Latin West, the first collector of Aristotle's study of animals and its Avicennian reworking was Albert the Great (d. 1280). Between 1258 and 1262, Albert composed his *Book of Animals* (*De animalibus libri*), which is a 26-volume commentary on Aristotle's zoological corpus. This commentary was part of his larger project of exposition of Aristotle's natural philosophy. Albert produced a continuous explanation of Aristotle's zoological writings, which combined a paraphrasis of Aristotle's *Book of Animals* (in the first 19 books) with doctrinal digressions, original expositions and inventories of animals useful for homiletic purposes or scriptural exegesis (Perfetti 2017, p. 291).

Embryology has remained the most disputed and lively part of Aristotle's zoological theory. Following Avicenna, Albert approached this issue by reconciling Aristotle and Galen. He distinguished three elements involved in sexual generation: male semen, female semen and menstrual blood. The first element has a fully active, formative power (*virtus formativa*). He then ascribed to the second element an informative power (*virtus informativa*), partly active and partly passive, which is entirely subordinate to the power of the male semen. Finally, he qualified the third element as completely passive because it contributes only the material principle to conception.

At the end of the 13th century, in the embryological treatise *On the Formation of the Human Body in Uterus* (*De formatione humani corporis in utero*), Giles of Rome (d. 1316) reaffirmed the primacy of Aristotle's hylomorphic theory of animal generation. Relying on the interpretation of Aristotle's theory provided in Averroes' *Colliget*, a medical work just translated into Latin, Giles sharply distinguished two principles of animal generation: the formal, active principle contained in the male semen and the material, passive principle supplied by the menstrual blood. Using logical-deductive arguments, he then denied any active, formal role to the female principle, even if subordinate to the role of the male semen. Vaginal secretion is thus a mere physical phenomenon that is unnecessary for the generative process (Martorelli Vico 2002).

Any embryological theory truthful to Aristotle's teaching endorses the following claims: the male is the only supplier of semen (one-semen theory); the semen is derived from the concoction of the blood (hematogenous theory); the male alone contributes form (via the semen), whereas the female contributes only matter (via the menstrual blood) (hylomorphic theory); the various bodily parts are formed in a definite order after conception (epigenetic theory). Such a theory is found in Dante's *Commedia* (*Purgatory* 25, pp. 34–78). According to Dante, male semen causes the purest part of the menstrual blood to coagulate. What is obtained is a first fetation, which is already alive to the extent that it has the capacity to nourish itself and to grow. As it grows, the fetus acquires the capacity to perceive. The rational soul, understood as distinctively human intellectual power, comes only at a later stage and from the outside. For Dante, this power is implanted in the embryo directly by God. The same account is also endorsed in the unfinished *Convivio* (IV, p. xxi 7), where Dante adds the following important clarification: as humans, we are all endowed with this power, but we do not possess the same amount of it. Some of us are better disposed to receive this power, which is a precondition to engage successfully in a happy life.

10.7. References

Alpina, T. (2022). Exercising impartiality to favour Aristotle: Avicenna and 'the accomplished anatomists' (*aṣḥāb al-tašrīḥ al-muḥaṣṣilūna*). *Arabic Sciences and Philosophy*, 32(2), 137–178.

Alpina, T. (2023). Between matter and form: Complexion (*mizāǧ*) as a keystone of Avicenna's scientific project. *Early Science and Medicine*, 28, 398–428.

Balme, D.M. (1987). The place of biology in Aristotle's philosophy. In *Philosophical Issues in Aristotle's Biology*, Gotthelf, A. and Lennox, J.G. (eds). Cambridge University Press, Cambridge, 9–20.

Campbell, G.L. (2014). Origins of life and origins of species. In *The Oxford Handbook of Animals in Classical Thought and Life*, Campbell, G.L. (ed.). Oxford University Press, Oxford, 233–247.

Cerami, C. and Falcon, A. (2014). Continuity and discontinuity in the Greek and Arabic reception Aristotle's study of animals. *Philosophia Antiquorum*, 8, 35–56.

Cerrito, A. (2023). *Albert the Great (c. 1193–1280) and the Configuration of the Embryo: Virtus Formativa*. Palgrave Macmillan, Cham.

Falcon, A. (2021). The reception of Aristotle's biology in late Antiquity and beyond. In *The Cambridge Companion to Aristotle's Biology*, Connell, S. (ed.). Cambridge University Press, Cambridge.

Falcon, A. (2024). *The Architecture of the Science of Living Beings: Aristotle and Theophrastus on the Study of Animals and Plants*. Cambridge University Press, Cambridge, 246–260.

Gotthelf, A. (2012a). Teleology and embryogenesis in Aristotle's generation of animals II 6. In *A Teleology, First Principles, and Scientific Method in Aristotle's Biology*. Gotthelf, A. (ed.). Oxford University Press, Oxford, 90–116.

Gotthelf, A. (2012b). Data-organization, classification, and kinds: The place of the *History of Animals* in Aristotle's biological enterprise. In *A Teleology, First Principles, and Scientific Method in Aristotle's Biology*, Gotthelf, A. (ed.). Oxford University Press, Oxford, 261–292.

Hadot, P. (1990). *Simplicius. Commentaire sur les Catégories*. Brill, Leiden.

Henry, D. (2016). The failure of evolutionary thinking in Antiquity. In *A Companion to Science, Technology, and Medicine in Ancient Greek and Rome*, Irby, G.I. (ed.). Blackwell and Wiley, Oxford, 313–321.

Kruk, R. (2003). La zoologie aristotélicienne. Tradition arabe. In *Dictionnaire des philosophes antiques. Supplément*, Goulet, R. (ed.). CNRS Editions, Paris, 329–334.

Lennox, J.G. (1996). Aristotle's biological development. In *Aristotle's Philosophical Development*, Wians, W.W. (ed.). Rawman and Littlefield, London, 229–248.

Martorelli Vico, R. (2002). *Medicina e filosofia. Per una storia dell'embriologia medievale nel XIII e XIV secolo*. Guerini e Associati, Napoli.

Musallam, B. (1987). Avicenna x. Medicine and biology. *Encyclopaedia Iranica*, III(1), 94–99.

van Oppenraay, A.M.I. (2017). Avicenna's Liber de animalibus ('Abbreviatio Avicennae'). Preliminaries and state of affairs. *Documenti e studi sulla tradizione filosofica medievale*, 28, 401–416.

Perfetti, S. (2017). La disseminazione del sapere sugli animali (dalla tarda antichità al XIII secolo) e l'iperaristotelismo di Alberto Magno. In *La zoologia di Aristotele e la sua ricezione dall'età ellenistica e romana alle culture medievali*, Sassi, M.M., Coda, E., Feola, G. (eds). Pisa University Press, Pisa, 269–297.

Primavesi, O. (2023). Review of Wellmann 2020. *Gnomon*, 95, 97–109.

Singer, P.N. (2021). Galen. In *Stanford Encyclopedia of Philosophy* [Online]. Available at: https://plato.stanford.edu/entries/galen.

Wellmann, T. (2020). *Die Entstehung der Welt. Studien zum Straßburger Empedokles-Papyrus*. Walter de Gruyter, Berlin.

11

Botany in Greece

Luciana REPICI
Università di Torino, Turin, Italy

11.1. In the beginning, Aristotle and Theophrastus

The two treatises by Theophrastus of Eresos (c. 372–288 BCE), *Enquiry into Plants* (*Historia plantarum*, nine books) and *On the Causes of Plants* (*De causis plantarum*, six books), can be considered unique in terms of ancient botany: they are the first writings we know of that are entirely devoted to the study of plants and the phenomena associated with them, and there is nothing in the written record that is comparable to this subject before or after this period[1]. Theophrastus was, however, a direct pupil of Aristotle (384–322 BCE) and, upon his death, took over the Peripatetic school where Aristotle had taught. The origin of the treatises that Theophrastus wrote should therefore be placed within the framework of Aristotelian thought and the extraordinary impetus given by Aristotle to the establishment of disciplines, such as zoology and botany, within his perspective of physics or the philosophy of nature. The two treatises also have a special role within this context, as they seem to represent the culmination of Aristotle's work on the study of living beings. Aristotle's work on plants is actually missing, alongside that devoted to animals. There are hints and references to the latter subject in the preserved

1 For recent translated and annotated editions of both treatises, see Einarson and Link (1976–1990) and Amigues (1993–2006, 2012–2017). A five-book work on plants is also attributed to another schoolfellow of Theophrastus, Phainias of Eresus. However, the condition of the work, fragmented and lacking in conceptual framing, does not allow us to draw any firm conclusions about the approach and content developed therein (see Anceschi 2015; Siede 2015; Zucker 2015).

History of Life Sciences and Medicine 1,
coordinated by Lorenzo PERILLI.

writings[2], but the work did not survive, perhaps because it was only planned and never written or because it was lost, and the work entitled *On Plants* (*De plantis*, two books), which has survived under his name is, in fact, as we shall see later, of questionable authenticity[3]. Reading Theophrastus' treatises is therefore recommended not only for their originality but also because, through their specific approach, they also reflect the interpretative lines adopted by Aristotle in his research on living nature[4].

Historia plantarum not only adopts the title of Aristotle's *Historia animalium* (*History of Animals*) but also the method and argumentative procedures, even extending to some areas of research that are necessary when looking into the nature of plants, which are not part of the research into the nature of animals. Both books are studies of the various differences found in the constitution and conformation of animals and plants, as well as in their way of being, reproducing and relating to the environment. There is also a similarity in terms of the argumentative procedures. Aristotle is accustomed to building his arguments by critically discussing facts, as well as highly reputed or common opinions, and using the widest possible sources of information to gather observational data useful for reasoning. This is also the case in his *History of Animals*, where information is acquired not only through direct observation but also from technical sources (farmers, fishermen, lumberjacks, etc.) and reports from indirect sources. In his *History of Plants*, Theophrastus uses similar procedures, referencing, assessing and discussing information from a variety of sources: poets, philosophers, plant specialists among his predecessors, accounts from explorers, as well as plant users at times, whose opinions verge on superstition and charlatanism. From a proposed methodological perspective, the study of a virtually infinite number of differences does not result in the establishment of a list or catalog; in the case of plants, these differences also include phenomena such as flowering and fruiting. Theophrastus' effort instead consists of disciplining and organizing a subject that does not have a classification, seemingly owing to its wide extent. Thus, a classification of plant parts is proposed and divided into different levels. The basic level consists of the parts on which life and death depend: moisture and heat; vital fluid or sap, fibers, veins and flesh; among these are the so-called

2 See, for example, *Historia animalum* (V 1, 539a21; *De Generatione animalium.* I 2, 716a1; I 23, 731a29).

3 There is, in fact, an account by Galen that mentions a work by Aristotle, which was closely linked to a treatise by Theophrastus on plants, but it was rare and difficult to find and was subsequently lost: see Gal. *De indolentia* 17 (p. 7, 7–11 Boudon-Millot and Jouanna 2010). However, as the editors point out (p. 70), it does appear to be authentic as there is a close link with Theophrastus' treatise, even though Galen copied it and does not indicate what its title was.

4 For a more detailed comparison, see Repici (2020a, pp. 13–63, 254–326).

homogeneous parts – "homoeomerous" in Aristotelian and Theophrastean terminology – in other words, wood, bark and pith, while the highest level corresponds to the so-called non-homogeneous parts – "non-homoeomerous" – which include root, stem, branches and sprouts. These are the organic parts of plants that carry out their own vital functions, such as feeding and growth, and are equivalent, albeit at a less complex stage, to the organs of animals[5]. Modeled on and adapted to the nature of plants, this classification, which is borrowed from the analogical classification of animal parts, introduces the widespread use of animal analogy as a heuristic tool in Theophrastus' botany. Although he is aware of the limits of its applicability (*Historia plantarum* I 1, 4), Theophrastus, similar to Aristotle, regards analogy as a useful classification criterion that enables objects characterized by different but similar functions to be compared and thus explains what is least obvious and unknown through what is most obvious and easily observable[6]. Theophrastus also provides a classification of the major plant genera, which were the first known to us, identifying four types: trees, shrubs, undershrubs and herbs. He does, however, warn that these definitions must be taken in a general and schematic sense, as the differences between these types are not obvious. The very nature of these objects studied may explain a certain lack of systematic rigor that he is sometimes criticized for in an anti-Aristotelian manner (Gotthelf 1988). According to Theophrastus, however, the starting point should be the definitions, as they contain characteristics that are common to all types and can therefore help to find the causes that are common to all. Indeed, to avoid time-consuming research, we must identify both common distinctions without making individual differences, as well as identify definitions and causes that are common to all types (*Historia plantarum* I 3, 2–6).

The key features of botanical knowledge in *Historia plantarum* therefore concern plant morphology, anatomy and physiology; in addition to phytopathology, plant diseases, whether they be reversible or fatal, are another area in which many differences can be detected (IV 14). There are also interpreters who consider Theophrastus an ecologist due to his attention to the determining impact that the natural environment can have on the existence of plants (Hughes 1988). However, it would be anachronistic and difficult to base the requirements of a discipline that emerged as autonomous in the second half of the 19th century, on the 4th century BCE. In fact, the study of what we now call environment is, for Theophrastus as for Aristotle, the study of external natural causes (essentially heat

5 See *Historia plantarum* I (1, 9– 2, 6) and on the problem (Repici 1985).

6 For Aristotle's classification of animal parts, see *Parts of Animals* II 1; for the identification of analogy as a criterion for classifying parts according to the parameter of the common function they fulfill in different genera, see *Parts of Animals* I (4, 644b11–23).

and cold in their various forms), which, by interacting with the internal factors of the individual constitution, can promote or destroy life, thereby indicating that living is a condition that requires a balance between the internal and the external factors, which, far from being obvious, can accidentally become difficult or uncomfortable[7]. However, plants are also the center around which a series of arts (*technai*) and pseudo-arts revolve, and according to Theophrastus, this aspect should not be overlooked in a treatise on botany. Readers of *Historia plantarum* will therefore find numerous sections on the use of wood for construction, flowers and plants for wreath-making, sowing plants for food, as well as juices and resins for the preparation of perfumes and for use in pharmacology[8]. This is not to say that the book should be seen as a manual for carpenters and woodworkers, flower growers, perfumers, doctors or pharmacists. In contrast, there are other purposes. The first is to extend the scope of noticeable differences to provide a solid inductive foundation. Another is the variety of properties that plants offer, confirming the extraordinary complexity of their constitution, despite the less organic nature of their structure. However, and above all, the question introduces the relationship between man and plants. Theophrastus makes no value judgment on the way in which humans understand and beneficially exploit the skills that they function with in this sphere. On the one hand, as in the case of the use of plants for the preparation of medicinal remedies, he points to deteriorating pseudo-arts such as those of root cutters and drugs or amulet sellers; on the other hand, as in the case of the properties of plants that affect bodily conditions and states of mind, he shows that this relationship can have unwanted implications if not all aspects of the plant's nature are known[9].

Agriculture also plays a part in this discussion on the basis of man's relationship with plants. However, it is not just because the subject is linked to human utility in terms of food that this skill takes on particular significance. In the case of plants, cultivation is a cause of generation that comes with nature and, like nature, is capable of generating living beings; in other words, capable, as Aristotle showed, of being born, feeding, growing and dying, such as animals and humans. Theophrastus therefore largely deals with agriculture in his other botanical treatise, *De causis plantarum*, which, much like the Aristotelian treatises *Parts of Animals*, *Generation of Animals* and other short monothematic writings on physiological subjects, adopts a causal method of explanation. In this vein, the first two books look at aspects of plant life, such as generation and germination, flowering and fruiting. The sphere of

7 These questions are examined by Aristotle in his short text *On the Longevity and Shortness of Life*, on which see Repici (2017, pp. 35–49, 111–120).

8 On this aspect, which concerns the relationship between nature and *techne* in the constitution of plants, see Repici (2020b).

9 On this question, see Preus (1988) and Repici (2006), with evidence and discussion of the reference texts.

reference here is nature, the sphere of causes on which plant life itself depends. Plants are subject to climatic variations and seasonal events that do not always ensure their survival, providing them with adequate and sufficient nourishment, as they are unable to defend themselves given their immobility. Unlike animals, in which reproduction has only natural causes, plants are also generated by human devices, and this is the fact that presents various problems for Theophrastus, who ponders on it in the following books (III–V) of *De causis plantarum*. The result is a very general classification that differentiates between plants that grow by nature or "spontaneously" because they do not depend on human labor and plants that grow owing to technical contrivances (*De causis plantarum* II 1, 1; III 1, 1), to which other classifications can be related such as status (domestic/wild), sex (male/female) and location (terrestrial/aquatic). However, this does not mean that nature and *techne* have the same status. Both Theophrastus and Aristotle recognize that they both operate by pursuing the good of the plant as an end; in other words, teleologically (*De causis plantarum* II 1, 1); their objectives, however, are different. Technical production is aimed at human utility, whereas nature is aimed at the generation and preservation of trees and plants for its own sake (I 16, 1–2). Human *techne* can also have the opposite effect, ultimately transforming plant nature for the worse if, through inexperience or ignorance, it fails to apply the right procedures. Theophrastus, therefore, carefully elaborates and analyzes the precise rules to be pursued and correct procedures to be applied. For, as Aristotle already argued, human skill requires experience and knowledge of causes; it is in those who love understanding, that is, philosophers, that understanding is rather reached through reason and causes only (III 2, 2–3). There are also objective limits to the action of human skill, and it is the very nature of plants that establishes them in all cases where they reject cultivation (III 1, 2–6). Priority is therefore given to nature, which is not inferior to *techne*, even in the production of seemingly astonishing things such as out-of-season fruiting or the production of fruit from unusual locations: the wonders of nature can astonish far beyond the wonders of human devices (III 1, 2–6; V 1, 2–4).

11.2. In the distant past and the near future

However, plants were studied even before Aristotle and Theophrastus by the "natural philosophers" of the old past – *physiologoi* according to the terminology adopted by both Aristotle and Theophrastus – in their research about the nature of the universe, although the precarious and fragmentary state of related reports does not facilitate complete reconstructions. As a result, plants were dealt with in a cosmological perspective, that is to say, as part of a general inquiry into the constitution of the world and its constituent entities, and this is the context in which the reflections of these more ancient authors on botanical questions should be placed. Theophrastus himself provides various indications of this effect, testifying to the breadth and concentration of the issues addressed in his treatises. We thus know that plant classifications were already being researched in the 5th century. Among Pythagoreans, Hippon of Rhegion dichotomized plants into wild and domesticated, fruit-bearing and non-fruit-bearing, flowering and non-flowering, evergreen and deciduous, identifying different causes in each case. It is the presence or absence of treatment, in other words, nature or man, that has caused the difference between the wild and the domestic world, but it is location and climate that have determined fruiting, flowering and perennial retention of leaves, or conditions contrary to these[10]. We are also informed that there was, too, the problem of spontaneous generation in plants, that is, where generation occurs without apparent cause. In line with his explanation of the origin of the world from an infinite number of principles known as seeds and the partial presence of these seeds in everything, Anaxagoras of Clazomenes (around 500–428 BCE) believed that the air contained the seeds of all things and that the latter were transported to the ground by rainwater; thus, plants are generated, although the seeds are invisible to the naked eye. In turn, Diogenes of Apollonia (second half of the 5th century BCE), who identified "intelligent air" as the sole principle of all things, believed that the process was due to the putrefaction of water and its mixing, to some extent, with earth. As for Cleidemos, a 5th-century botanist known only to us through Theophrastus, he thought the cause should be attributed to the characteristics of the nature of the plants themselves, which, unlike those of animals, are much more turbid and colder[11]. Therefore, we can assume that, at the time, the analogy between animals and plants was somehow already working. Theophrastus also credits Cleidemos with taking a stand on another problem that Theophrastus himself took issue with, namely, what are the sensible effects produced in plants by heat and cold, from which it is possible to define whether the nature of a plant is hot or cold? Cleidemos took as a starting point the fact that annual plants do not bear fruit at the same time of year but in different or opposite seasons, as well as the fact that they ripen sooner or later. In his view, this was due to the balance between the hot or cold nature of plants and the opposing properties

10 *Historia plantarum* I 3, 5; III 2, 2, in Hippon (38) A 19 Diels/Kranz. In the sources, Hippon's place of origin is also indicated as Metapontum and Samos.

11 See Theophr. *Historia plantarum* III 1, 4–5; *De causis plantarum* I 5, 1–5, in Anaxagoras (59) A 117, Diogenes (64) A 32, Cleidemos (62) 3 Diels/Kranz. Cleidemos may also have been working on questions of agricultural technique, as Theophrastus relates certain works to him, such as instructions on sowing during the sunset of the Pleiades and the explanation of certain pathological anomalies in plants, such as excess foliage in vines due to excess water (*De causis plantarum* III 23, 1–2, in Cleidemos (62) 5; *De causis plantarum* V 9, 10, in Cleidemos (62) 6 Diels/Kranz).

of the season: cold plants bear fruit in summer, and hot plants bear fruit in winter, since the action of the opposite avoids any excess in either direction that would prevent fruiting[12]. On the basis of this same principle of the action of opposites against each other and the resulting balance, a whole series of evidence is provided by Menestor of Sybaris, a Pythagorean, also from the middle of the 5th century, who was also interested in problems of botanical nomenclature[13]. Thus, the indications of a cold nature were sterility, and those of a warm nature, fertility; adaptation to warm soils indicated a cold nature, that to cold soils a warm nature; and the symptoms of a warm nature were early germination and fruiting, and the latter were also evergreen plants, whereas it was the absence of heat that made them deciduous. In addition, Menestor added case evidence of combustible plants such as ivy, whose heat melted even snow[14].

Menestor's theses were shared by Empedocles of Acragas (c. 483–423 BCE), and both argued that the same principle applied to both plants and animals; in other words, warm natures require cold places, such that the colder they are, the warmer their nature is, and vice versa, which explains why there are plants capable of living in water, similar to animals[15]. However, in the specific case of plants with "perennial foliage", Empedocles believed that the phenomenon instead depended on a particularly mild outside air temperature and a porous constitution on the inside, allowing easy absorption of nutrients[16]. The relationship between plants and their place of growth is thus related to another aspect: the importance of nutrition for their very existence. This point is also recognized by others but explained differently. For Empedocles, there are pores or channels in the structure of plants, the size of which would be proportional to that of the nutrition flowing in from the outside, allowing them to pass through and spread. Anaxagoras, as we know from other sources, was also interested in this issue. However, in his explanation, the case of plant nutrition is emblematic of the principle of the partial presence of seeds of all things in

12 *De causis plantarum* I 10, 3, in Cleidémos (62) 4 Diels/Kranz.

13 According to Theophrastus, Menestor called plant moisture *juice* (*Historia plantarum* I 2, 3, in Menestor (32) 2 Diels/Kranz).

14 *De causis plantarum* I 21, 5–7, in Menestor (32) 5 Diels/Kranz. See also *Historia plantarum* V 3, 4; V 9, 6. In the case of Menestor too, we can assume that there was an interest in agricultural technique. Theophrastus relates work to him, notably instructions on the use of oily earth for fertilization, given its drying effects on plants: *De causis plantarum* II 4, 3.

15 Ivi, I 21, 5, in Empedocles (31) A 73 Diels-Kranz. This information can already be found in Aristotle's *De Respiratione*, 14, 477 a32.

16 Ivi, I 13, 2, in Empedocles (31) B 77–78 Diels-Kranz. In context, Theophrastus also recalls the words of the poets on this subject. See *Odyssey*, VII 114-121.

everything ("everything is in everything"): in the water and food that plants draw from the earth, there are invisible seeds that constitute all of their parts, and it is for this reason that they grow in the form of roots, stems, branches or leaves; in other words, with differentiated components[17]. Theophrastus, in turn, relates the opinion of Democritus of Abdera (born approximately 460 BCE), the main representative of ancient atomism, with his master Leucippus. For Democritus, the nutrition of plants is characteristic of the growth of atomic compounds. The pores contained in the structure of plants are, in fact, interstitial voids that, as body components such as atoms, allow the passage of matter from food and its distribution to the whole. In plants, early or late germination and fruiting, as well as greater or lesser longevity, depend on conformation, which is determined by the way in which atoms aggregate. In straight-stem plants, food spreads rapidly, allowing early germination and fruiting, but the plant is consumed and perishes early; the opposite is true of twisted-stalk plants[18]. Using the principles of atomic physics, Democritus also explained the formation of aromas or juices in plants. He identified several basic types of flavor (sweet, sour, salty and bitter) and associated certain atomic shapes with each of them: sweet, for example, had round-shaped atoms of a certain size; bitter, however, had round-shaped, smooth and hooked atoms but were small in size. This not only makes it possible to explain the atomic nature of the object (juice) but also to show that the subjective view of it depends on the conformation of the object: the qualities of a body that is sweet or bitter are not the main properties but also derive from the properties of the atomic components and the form that the atoms take on in the constitution of each aggregate[19]. Menestor was of a different opinion. According to him, the juices/tastes of plants were in fact innumerable, as they resulted from their mixing and blending with one another and differing in taste according to how much of one was mixed with the other[20].

There is, however, one point on which Anaxagoras, Democritus and Empedocles agree: the question of the animation of plants, in other words, whether they are endowed with psychic abilities, such as those of animals and humans. The evidence of the accord from the three philosophers regarding this point does not come from Theophrastus himself, who actually neglects dealing with the question of plants' *psyche* in a systematic way, but from the Peripatetic tradition in the pseudo-Aristotelian work *De plantis*; for Anaxagoras, as for Democritus and Empedocles, plants not only have desires and feelings such as joy and sadness but also intelligence and the ability to understand. Anaxagoras finds evidence of this in

17 See Simpl. in *Aristotle*, Phys. 460, 4, in Anaxagoras (59) A 45 Diels/Kranz.

18 *De causis plantarum* II 11, 7–8; I 8, 2, in Democritus (68) A 162 Diels/Kranz.

19 *De causis plantarum* VI 1, 6–7, in Democritus (68) A 129 Diels/Kranz.

20 *De causis plantarum* VI 3, 5, in Menestor (32) 7 Diels/Kranz.

the deciduous nature of plants: it is a sign of sadness that they lose their leaves[21]. The underlying hypothesis in all three cases is the common formation of all things from the same principles (Anaxagora's infinite seeds, Democritus' infinite atoms, Empedocles' four elements or "roots"). Therefore, there is nothing to rule out the possibility that, within the living world, plants may have the same psychic traits as animals and humans. Thus, the analogy between plants and animals is more complex, and the problem begs the question of whether there is a delineation between these different conditions of life. In line with their respective theories, Anaxagoras and Democritus also asserted that the plant is an "animal planted in the earth"[22]. This was only the beginning of the discussion. Plato, who saw a form of irrational desire for food and drink in plants, compared man to a "celestial plant". By this, he meant that man's roots – that is to say, the part that constitutes the true essence of the human being, the intelligence or rational soul that cannot be found in plants – were located in the sky, the direction in which his own upright stature leads[23]. It was Aristotle who overturned this Platonic proposition. In his view, plants, animals and men were united by the possession of a nurturing soul; in other words, the basic function of all living beings, that of nourishment and growth. However, particularly in the relationship between plants and humans, the organ responsible in both cases for performing this function – the roots in plants, the head and mouth in humans – is located at opposite ends in space: below in plants and above in humans. We could therefore say that plants are like upside-down humans[24].

This theme seems to have been the focus of the philosophies of the post-Aristotelian era. For Stoics and Epicureans, plants were not animated beings but were inanimate in nature. What made them so was the absence of an internal principle of movement (namely, the soul), which was at the origin of impulsive or desiring motions, such as in animals and humans, as well as at the origin of acts of thought in humans alone. Nutrition itself makes them living beings but not animate ones: it is not the vegetative aspect that makes an animal an animal and a man a man but rather a sensation and reason[25]. Therefore, there is a clear line separating plants from animals and humans, and no analogy can be made here, except by metaphor. Thus, they are part of nature, and their mode of life, generation, growth and reproduction, as well as all related phenomena, are governed by nature itself.

21 *De plantis* I 1–2, in Anaxagoras (59) A 117, Democritus (68) A 163, Empedocles (31) A 70 Diels/Kranz.

22 See Plut. *Quaestiones Naturales* 1, 911d, in Anaxagoras (59) A 116 Diels/Kranz.

23 See Plato, *Timaeus* 77a–c; 90a–c, and Repici (2020a, pp. 231–241).

24 On this question and the relevant Aristotelian reference texts, see Repici (2020a, pp. 27–31).

25 In *Stoicorum Veterum Fragmenta* II 708-709, both positions are reported together.

With respect to their way of understanding nature, Stoics and Epicureans diverge: in Stoic nature, the regulating principle of plant life is the universal divine reason that governs everything as providence, whereas in Epicurean nature, the being of all things is regulated by the movement of atoms in the void and their random interweaving[26]. Botany, however, as a discipline in its own right in the broader context of the philosophy of nature, has no special place within the general philosophical schemes of Hellenistic schools, and there is no mention of specialized botanists, even among the intellectuals and scientists who, from the early 3rd century BCE, convened around the museum and library of Alexandria.

11.3. In Rome

Theophrastus' original investigation on botanical matters, then, does not seem to have had any heirs or successors, and the fact that its disappearance is recorded in the Peripatos itself, such as that of Aristotle's zoology, is generally interpreted as a sign of the school's decline, which can be attributed to a loss of interest in theoretical approaches in favor of learned erudition[27]. However, the school's decline, if any, may have lied in its difficulty in acquiring interest in building a botanical philosophy, as well as competition from botanical medicine, which focused on the study of plants in terms of their pharmacological utility. The success of a treatise on medical herbalism such as Dioscorides' *De materia medica* (1st century CE) is emblematic of this disparity. For Dioscorides, it was not just a matter of finding out what the natural powers of plants were but also of determining their preparation and dosage from a therapeutic perspective and even methods of preservation[28]. The application of theoretical botany to agriculture, which Theophrastus had discussed at length, was difficult for those cultivating a field, as M. Terentius Varro (116–127 BCE) pointed out among the authors of Roman treatises on agronomy. It seems that Theophrastus' work was not used in Arab treatises on agronomy either[29]. In Pliny the Elder's *Naturalis historia* (23–79 CE), Theophrastus is cited as one of the authors he followed in writing books on plants, and there are several references to

26 For a more detailed analysis, see Repici (2010, 2015, pp. 13–28).

27 For discussions on this issue, see Sharples (1999) and Lennox (2001).

28 On the organization of Dioscorides' work, its methods and objectives (Touwaide 1993).

29 See Gutas (1985) as well as Daiber (1985, pp. 103–104), who reports that Theophrastus and his *De causis plantarum* are quoted in Ibn al-Nadîm's *Fihrist [Index]*, completed around 987, where we are told that the work had been "translated by Ibrâhîm b. Bakkûs. There is only a commentary on certain parts of the first book". However, at the end of his analysis, Daiber (1985, p. 108) concludes that almost certainly, Theophrastean botany was not known in the Arab world, and that it was instead taken from the pseudo-Aristotelian work *De plantis*, or even from Dioscorides and Galen, but not directly from Theophrastus.

his two botanical treatises[30]. However, *Naturalis historia* is not a work on theoretical botany; in contrast, as we know, it is an encyclopedic survey that includes all of the kingdoms of nature, and although there are 15 books on plants – 10 of which (XII–XXII) are methodical studies and five (XXIII–XXVII) are on medicinal herbalism – they only mention one facet of the great spectacle of nature that Pliny so admires. A large amount of data and information has been collected from various disciplines in addition to botany: astronomy, meteorology, geography, anthropology and psychology, zoology, pharmacology, mineralogy and metallurgy. Botany itself brings together arboriculture, gardening, floriculture and even indications of how to run a farm (books XVII–XXI)[31]. This does not, however, lead to disruption and indiscriminate accumulation. In fact, the more or less broad dichotomous nature of the classifications, between the extreme differences in the various cases, provides a direction for this field. As a whole, Pliny's work can be considered a study of the differences through similarity or dissimilarity, such as Aristotle's *Historia animalium* and Theophrastus' *Historia plantarum*, with due differences. Thus, books XII–XVII focus on the broad classification between spontaneous, that is, "natural", plants and cultivated plants, particularly for arboricultural plants (trees and shrubs). The specific context of books XII–XIII is the distinction between exotic plants, originating from foreign territories, and the more familiar, common plants, which are typical of nearby Latin territories, constituting an application of the classification of plants according to their place of origin and location. A choice of method could also be perceived in the initial treatment of trees, justified by the fact that men could use them if they were not blinded by artificial needs, as well as by the reminder of old beliefs still present among simple country folk, that trees are the temples of the gods[32]. The moral denunciation that condemns the abuse of trees and plants, be they exotic or otherwise, in the name of luxury, such as in the vigorous search for *felix ac beata Arabia* perfumes – which is just a name and imagination – is also part of the overall picture[33]. There is also an internal classification in the background of books XV–XVI, which is specific to non-cultivated plants and is

30 See, for example, the long discussion in *Naturalis historia* XVI 19–34 of the case of oaks (cf. Theophrastus, *Historia plantarum* III 7, 4 et seq.; III 8, 4; III 16, 2).

31 According to Healy (1999), the significance of the work specifically lies in the fact that it provides an encyclopedic framework for scientific and technical knowledge in the 1st century CE. For other interpreters, on the other hand (Conte 1982; Beagon 1992), Pliny's encyclopedia is not just a compilation, but a work with practical purpose, and should be studied as a whole, rather than with reference to the individual disciplines that are mentioned in it.

32 *Naturalis historia* XII 1–8; XII 13.

33 *Naturalis historia* XII 51–71; XII 82; XII 113 and see Theophrastus *Historia plantarum* IV 4, 14; IX 6, 2; IX 6, 1.

applied to distinguish between tree species with domestic fruits, such as apples, grapes and figs, and plants with wild fruits, such as acorns. There is an obvious underlying distinction between trees with edible fruit and trees with non-edible fruit, but it can itself be reduced to a more general distinction in terms of wild plants (XV 138), between places where no trees grow, such as deserts, and places where many trees grow (XVI 1–6). Differences in conformation, use and geographical location, sex and taste of the fruit are also reported among trees of the acorn genus (XVI 7–17)[34]. However, from an almost humanized perspective of plants, Pliny does not overlook the differences linked to the cultivation sites "preferred" by plants, which condition height, beauty and better fruiting (XVI 77). Similarly, in establishing the distinction between wild and domestic plants, on which the distinction between deciduous and evergreen plants depends, Pliny characterizes domestic plants as being "more civilized" (*urbaniores*), defined by a greater "gentleness" than others – as in the case of the human *societas* – and therefore capable of performing the task of benefiting mankind "with more humanity" (XVI 78).

The most important and fundamental difference, however, is identified in the modes of plant generation: nature and human skill and contrivance, as with Aristotle and Theophrastus. There are three natural plant generation methods: spontaneous, from seeds and from roots, while many others are linked to technical intervention. However, nature and *techne* cannot simply ignore the necessary causes of external environmental factors in their purpose-driven efforts, and both must accommodate the internal constitutive nature of plants. The plants' adaptation or non-adaptation to a certain environment may depend on the latter, both when they grow naturally, in other words, without technical care, and when they are cultivated (XVI 134). Here, we identify one of the theoretical constructs that underpins the structure of *Naturalis historia*: the problem of why plants exist. For Pliny, nature and technical ability are both causes of plant existence but not at the same level. The superiority of nature over *techne* is indisputable: the inventiveness and ingenuity of technical operations may be astonishing, but they do not have the completeness, perfection and power of nature, nor the extraordinary precision and imagination the latter reveals in the production of flowers and scents; in other words, products not even intended for human use. It is nature, not agriculture, that provides food for mankind, and it is nature that provides medicines; hence, Pliny's well-known denunciation against the "frauds" of medical art, particularly the Greeks, accused of deception such as magic[35]. Nature also cannot be mimicked; not only was it unrealistic to believe that

34 Acknowledging the sexual distinction between plants, which is not recognized by Aristotle, Pliny follows Theophrastus: see *Historia plantarum* III 8, 1; *De causis plantarum* I 16, 5; II 10, 1–3.

35 *Naturalis historia* XXIV 4–5; XXIX 1-28; XXX 1–18.

it could be reproduced in its original state in villa gardens, which were no more than luxury exhibitions, but it was also unnatural to reproduce it pictorially in landscape scenery or in herbariums, such as that of Crateua. For Pliny, however, rendering the liveliness and variety of nature's colors in images and figures was not just wishful thinking; copyists also disservice through their negligence. In his opinion, images and figures primarily immobilize nature because they draw plants as they are at a certain time of year, without considering the fact that their appearance varies with each season (XXV 8–9). Plants are living beings, therefore subject to evolution, and as such, they are not soulless either: "nothing lives without a soul", writes Pliny, taking a stand on the ancient debate (XII 1). It is the annual order of nature itself that marks the stages of their lives, described as some sort of an earthly origin. First, at the conception stage, it is the warm foehn spring breeze that favors generation; Pliny presents this *ventus* in Stoic terms as a *genitalis spiritus mundi;* in other words, as the generative breath of the world, which gives the earth a kind of burning desire to receive seeds and "give life (soul) to all that is sown". What later corresponds to generation and birth in animals is germination and flowering in plants, whereas fruiting is understood more as some sort of breeding (*educatio*). This is what Pliny calls the *labor arborum*, that is to say, the completion of the growth process, where the reference to labor seems to underline that the process is not achieved without a certain effort and difficulty in its realization. However, at least in the case of trees and plants that naturally bloom, fatigue is seen as a kind of compensation for the joy (*gaudium*) of flowers (XVI 94–111). The "benevolent earth" and "mother earth" look after them with care, relieving the sadness that flowerless trees and plants have[36].

11.4. The missing link

Around the year 1200, Englishman Alfred de Sareshel translated from Arabic into Latin a work *On Plants* and presented it as a work by Aristotle. There appears to have been an earlier Latin translation, predating a Syriac version that was taken directly from a Greek original, which an Arabic translation was added to, then later rendered into Latin by Alfred, and finally an anonymous Greek retroversion. It is this two-book retroversion of *De plantis* that we are reading today: a work lost in its original Greek version, which has undergone five translations into different languages (Syriac, Hebrew, Arabic, Latin and Greek) and exposed to all of the remakes and alterations easy to imagine in the passages from one language to another. Moreover, in Near Eastern regions, the author of this work is referred to as the peripatetic Nicolaus of Damascus (born approximately 64 and supposedly died

36 *Naturalis historia* XXIV 1; XVIII 1–5; XXII 1; XXIV 2. For a general overview of Pliny's work in ancient botany and the questions here dealt with, see Repici (2015, pp. 135–152).

in Rome at an unspecified date after 4 BCE), known in the West for his compendium of Aristotelian philosophy and for his interest in the philosophy of nature[37]. These reasons alone would be enough to make us doubt that this is an authentically Aristotelian work. However, the attribution to Aristotle made up for the absence of a work of his on plants, which could be seen as a gap in his output of studies on the philosophy of living things. From then on, *De plantis* became the manifesto of Aristotelian botany. Actually, the work combines elements from both Theophrastus' treatises and Aristotle's reflections, and thus establishes a compilation without any original thought. Furthermore, perhaps as a result of manipulation over the course of its history, the text is inaccurate in its terminology and fragmented in its structure. There are frequent gaps and repetitions in the development of the discourse, neglecting clarity and often comprehensibility, and especially in the second book, the presentation takes the form of a collection of data with no apparent order or systematic structure. It is also difficult to tell what the intention of the work's presumed author, Nicolas de Damas, was (to provide Peripatos with a basic compendium of the discipline? To merge Aristotle's hypotheses on the subject and Theophrastus' development of the discipline into a single text on Aristotelian-Peripatetic botany?), or what audience it was intended for (schoolchildren? External non-specialists?). However, even considering its severe constraints and uncertainties, *De plantis* represents a significant moment in the history of ancient botany, as it documents the effort made to give the discipline an autonomous theoretical framework, not applied to specific fields such as medicine, and instead adopts a method of treatment on the basis of reasoning and observational data, as implemented by Aristotle and Theophrastus.

As a whole, with a continuation between the first and second books, and leaving aside aspects of empirical dispersion, the work is structured around a number of key themes.

1) Plant animation (Book I): The historical-critical comparison with the theses of the most ancient philosophers, which have already been referenced, is also part of the discourse. The author of *De plantis* opposes their positions, assuming that plants cannot have the psychic characteristics of animal living beings (sensation and/or thought) or the organic functions that the latter have – respiration, sleep/wake alternation and sexual distinction – only nutrition, growth and reproduction. Plants only have, what Aristotle calls, the nutritive soul.

2) Plant conformation: The analogy with animals between similarities and dissimilarities is also a key theme. The detection of the variety, inconsistency and

37 On this topic and related discussions of the history of studies, see Repici (2009). The standard edition of the work is edited by Drossaart-Lulofs and Poortman (1989). For an Italian translation with a detailed introduction and valuable comments, see Ferrini (2012). On the issue of the authenticity of the Aristotelian work, see section 11.1 and footnote 3 above.

dissimilarity of parts that plants are made of is accompanied by signs of their distribution in different groups, as well as of the correlation between body parts and their reproductive capacity. The types of parts identified are, using Aristotelian and Theophrastsian terminology, "homoeomerous" and "non-homoeomerous" and, like the two philosophers, the author of *De plantis* also notes a lack of organic components in plants and thus of specialized functions and complex purposes. In this context, a classification of plants according to genus is applied, ranging from a particular genus to a general genus on an inductive basis.

3) Plant constitution and nutrition (Book II): By conformation, plants are simple, non-articulated in nature, but in terms of constitution, they are cold and dry and are barely tempered by a certain degree of humidity and heat. The thin, rarefied (equals not thick or dense by nature) parts of their bodies are the source of their nutrition and growth: through them, food is absorbed by attraction from the earth and transformed into vapor through the act of internal heat, spreading rapidly upward to all parts of the plant, supporting its erect posture.

4) Plants and external nature: There are four environmental factors that determine growth: the right seed, an appropriate location, moderate amounts of water and temperate air. This explains the variations in structure and the different properties that plants and plant parts can take on, according to the characteristics and properties of their location and the climate.

5) Abnormal and unusual plants: A lack of nourishment or inadequate absorption in the external or inner nature of plants can cause degenerative phenomena and abnormalities, such as excess foliage or malformed growth. The method of investigation uses a problem-based procedure ("why *x*?"), which is typical of many writings in the Aristotelian tradition, while the causal framework mainly appears to be based on a mechanistic approach.

The fact that Alfred found it necessary to accompany his translation with a commentary already shows that, in his eyes, Aristotle's *De plantis* had not achieved a sufficient level of clarity. Albert Magnus (1206–1280) also noted obscurities and proposed fixing them. However, he did not associate these obscurities with Aristotle himself, who he too considered the author of *De plantis*, but rather with the linguistic and conceptual negligence, or misunderstandings, of his translators. With this strong declaration of principle in *De vegetabilibus* (I Treatise I 9–14), Albert undertakes a drastic revision of the text translated by Alfred, before presenting the subject in the name of Aristotle, which appears almost as a rewriting of a treatise on botany that Aristotle should or could have written it. Among the seven books that make up the work, in which Albert shares his own thoughts on the philosophy of living, there is also an herbarium. The sixth book is, in fact, a kind of catalog that, in alphabetical order, records examples of trees (Treatise I) and herbs (Treatise II) that have either been directly observed or taken from other authors, describing in detail their properties, various uses and often even their morphological arrangement. However, the insertion is primarily intended to satisfy student curiosity rather than philosophy, since, as he observes, philosophy is not compatible with details (Treatise I 1, pp. 339–340)[38].

Overall, Albert's work is also a treatise on botanical theory; in other words, the philosophy of plant nature, but it is simultaneously an original rewriting. As for the methodological criteria underpinning it, in addition to the need for a clear, fluid and uncontracted presentation, there is also a need to bring order to a complex subject that is difficult to discipline, proceed with a necessary systematic structure to avoid dispersion and establish general research principles before moving onto examine specific questions.

1) The nature of plants: In the treatment of the science of living beings, it is the nature of plants that must be studied first, as this is the level at which life is separated from sentience and presents less complexity than animal nature (Treatise I 1, paragraph 3).

2) Plant animation: In the order of problems, we must first look at plant souls; second, the powers of the plant soul that are unique to each of them, and whether they also have desires and sensations; third, the possible presence of sleep and awakening in plants; fourth, their sex; fifth, the degree of perfection of the plant in relation to the animal; and, sixth, their mode of life (Treatise I, paragraph 59). The conclusion is that plants are nurturing; in other words, they have a nurturing soul but do not desire, perceive, sleep or exhibit sexual distinctions and are therefore imperfect compared with the animal condition.

3) Plant parts and genera: Anatomical conformation varies greatly in relation to the disparity of parts that are present or absent. Nevertheless, it is possible to establish a general classification that distinguishes the following: characterizing parts, such as sap and root; "material" parts, such as nutrient juice; organic parts, such as nodes, veins, bark and roots, to be studied by comparison with animal parts; and variable parts, such as leaves, flowers, fruits and seeds. The morphological structure is therefore also very different with respect to factors such as ascending or descending growth or the shape of fruits and seeds (Treatise II. I 2–4).

38 The work is part of a renewed interest in the philosophy of nature in academic and Dominican circles and is part of Albert's project to build a universal science. For the positioning of this work and its significance for attitudes toward the life sciences, see Bianchi (1997) and the essays by Työrinoja et al. (1990) and Weisheipl (1994). To learn more about the originality of *De vegetabilibus* in the evolution of the *scientia de plantis* in the Middle Ages, see Panarelli (2020).

A classification into six major genera (trees, shrubs, fruits, vegetables, herbs and mushrooms) is also possible (Treatise II, paragraph 151); however, there are variations according to cultivation methods and growing locations, and there are a number of secondary species.

4) Plant generation and its causes: In accordance with their nature, plants live and exercise life by growing, feeding and reproducing. This being said, the reason for their being is the intermingling of celestial forces (as the masculine principle of generation) and terrestrial forces of birth (as the feminine principle of generation); in other words, as the ancient philosophers put it between father sun and mother earth (II Treatise I 1, paragraph 5–7). However, their generation is faster than that of animals, both because they are simpler in constitution and therefore do not require specialized nourishment and because their structure is made up of hidden pores through which food is channeled[39]. They can therefore be generated from a seed, or even without a seed and instead from putrefied matter, provided that heat, humidity and external climatic factors all reach the right conditions. The earth is their belly, and the roots are the mouth through which they absorb food (II Treatise I 1, paragraph 13). Aristotle derived his presentation of plants as upside-down humans from this; Albert remembers it but attributes it to Plato[40].

5) Plant nutrition: As a result of the simplicity of their elemental composition and their organic make-up, the process of plant nutrition is not a particularly complex mechanism. The nourishment drawn from the earth, of which plants are composed together with water and fire, is absorbed through the hidden pores of their structure, cooked by heat, and under its action it spreads to the upper parts. However, if we consider a plant's dependence on external natural factors, the influence of the location where it grows can be decisive (IV Treatise II, paragraphs 42–84). Nevertheless, the operations and technical arrangements that have been implemented to make wild plants domesticated for human use also count (VII Treatise II, paragraphs 126–184).

6) Accidental plant properties: These include tastes, smells, colors and even medicinal powers, all of which are characterized by variability. Here, the technical operations involved in domesticating plants are also important, and in this context, a descriptive procedure is crucial. However, dispersion can be avoided by knowing that for each of these properties, fixed classifications are given according to types, which are specific variations and differences, and by knowing the common cause of

39 II Treatise I, paragraph 1, 11–13; IV Treatise I, paragraph 12–14.

40 II Treatise I 5 paragraph 67: "planta dicitur a Plato esse figurae hominis conversi". In Treatise I 10, paragraph 67, the expression "plantam esse conversum animal" appears, but, as in his commentary on Aristotle's *De anima* (II Treatise I 3), Albert does not directly link it to Aristotle.

all of these properties, namely, the degree and completeness with which food drawn from the earth is absorbed, digested and fully processed.

11.5. References

Amigues, S. (1993–2006). *Théophraste. Recherches sur les plantes*, 5 volumes. Les Belles Lettres, Paris.

Amigues, S. (2012–2017). *Théophraste. Les causes des phénomènes végétaux*, 3 volumes. Les Belles Lettres, Paris.

Anceschi, B. (2015). The metaphor as a scientific device in the botanical description of the mallow in fragment 49 of Phainias of Eresus. In *Phaenias of Eresus. Text, Translation, and Discussion*, Hellmann, O. and Mirhady, D. (eds). Transaction Publishers, New Brunswick/London, 365–375.

Asper, M. (2015). Peripatetic forms of writings: A system-theory approach. In *Phaenias of Eresus. Text, Translation, and Discussion*, Hellmann, O. and Mirhady, D. (eds). Transaction Publishers, New Brunswick/London, 407–432.

Beagon, M. (1992). *Roman Nature. The Thought of Pliny the Elder*. Clarendon Press, Oxford.

Bianchi, L. (1997). *La filosofia delle università. Secoli XIII-XIV*. La Nuova Italia, Florence.

Boudon-Millot, V. and Jouanna, J. (2010), *Galien. Ne pas se chagriner*. Les Belles Lettres, Paris.

Conte, G.B. (1982). L'inventario del mondo. Ordine e linguaggio nell'opera di Plinio il Vecchio. In *Gaio Plinio Secondo, Storia naturale*, volume I, XVII-XLVII, Conte, G.B., Barchiesi, A., Panucci, C. (eds). Einaudi, Turin.

Daiber, H. (1985). A survey of Theophrastean texts and ideas. In *Theophrastus of Eresos: On His Life and Work*, Fortenbaugh, W.W., Huby, P.M., Long, A.A. (eds). Transaction Books, New Brunswick/Oxford, 103–114.

Drossaart-Lulofs, H.J. and Poortman, E.L.J. (1989). *Nicolaus Damascenus De Plantis. Five Translations. Aristoteles Semitico-Latinus*, volume 4. North-Holland Publishing Company, Amsterdam/Oxford/New York.

Einarson, B. and Link, G.K.K. (1976–1990). *Theophrastus. De causis plantarum*, 3 volumes. Harvard University Press, Cambridge/London.

Ferrini, M.F. (2012). *Aristotele. Le piante*, Bompiani, Milan.

Gotthelf, A. (1988). Historiae I: plantarum et animalium. In *Theophrastean Studies: On Natural Science, Physics and Metaphysics, Ethics, Religion and Rhetoric*, Fortenbaugh, W.W. and Sharples, R.W. (eds). Transaction Books, New Brunswick/London, 100–135.

Gutas, D. (1985). The life, works, and sayings of Theophrastus in the Arabic tradition. In *Theophrastus of Eresos: On His Life and Work*, Fortenbaugh, W.W., Huby, P.M., Long, A.A. (eds). Transaction Books, New Brunswick/Oxford, 63–102.

Healy, J.F. (1999). *Pliny the Elder on Science and Technology*. Oxford University Press, Oxford.

Hughes, J.D. (1988). Theophrastus as ecologist. In *Theophrastean Studies: On Natural Science, Physics and Metaphysics, Ethics, Religion and Rhetoric*, Fortenbaugh, W.W. and Sharples, R.W. (eds). Transaction Books, New Brunswick/London, 67–75.

Lennox, J. (ed.) (2001). The disappearance of Aristotle's biology. A Hellenistic mystery. In *Aristotle's Philosophy of Biology. Studies in Origins of Life Science*. Cambridge University Press, Cambridge, 110–125.

Meyer, G. and Zimmermann, A. (1980). *Albert Magnus Doctor Universalis 1280–1980*. Matthias-Grünewald Verlag, Mainz.

Panarelli, M. (2020). Albert the Great's *De vegetabilibus* and its unique position among the medieval commentaries on *De plantis*. In *PERI PHYTON. Trattati greci di botanica in Occidente e in Oriente*, Ferrini, M.F. and Giglioni, G. (eds). Edizioni Università di Macerata, Macerata, 137–162.

Preus, A. (1988). Drugs and psychic states in Theophrastus' Historia plantarum 9, 8–20. In *Theophrastean Studies: On Natural Science, Physics and Metaphysics, Ethics, Religion and Rhetoric*, Fortenbaugh, W.W. and Sharples, R.W. (eds). Transaction Books, New Brunswick/London, 76–99.

Repici, L. (1985). Il paradigma animale nella botanica di Teofrasto. *Rivista di Filosofia*, 76(3), 367–398.

Repici, L. (2006). Medici e botanica popolare. In *Medicina e società nel mondo antico, Proceedings of the Udine Conference (4–5 October 2005)*, Marcone, A. (ed.). Le Monnier, Florence, 72–90.

Repici, L. (2009). Il *De plantis* pseudo-aristotelico nella tradizione antica e medievale. In *Le monde végétal. Médicine, botanique, symbolique*, Paravicini Bagliani, A. (ed.). Edizioni del Galluzzo, Florence, 77–94.

Repici, L. (2010). Le jardin des *Parva naturalia* : les plantes chez Aristote et après lui. In *Les Parva naturalia d'Aristote. Fortune antique et médiévale*, Grellard, C. and Morel, P.-M. (eds). Publications de la Sorbonne, Paris, 31–45.

Repici, L. (2015). *Nature silenziose. Le piante nel pensiero ellenistico e romano*. Il Mulino, Bologna.

Repici, L. (2017). *Aristotele. La fiamma nel cuore: Lunghezza e brevità della vita, Gioventù e vecchiaia, La respirazione, La vita e la morte*. Edizioni della Normale, Pisa.

Repici, L. (2020a). *Uomini capovolti. Le piante nel pensiero dei Greci*, 2nd edition. Edizioni della Normale, Pisa.

Repici, L. (2020b). La botanica di Teofrasto e il mondo delle tecniche. In *PERI PHYTON. Trattati greci di botanica in Occidente e in Oriente*, Ferrini, M.F. and Giglioni, G. (eds). Edizioni Università di Macerata, Macerata, 75–106.

Sharples, R. (1999). The peripatetic school. In *From Aristotle to Augustine*, Furley, D.J. (ed.). Routledge, London, 147–187.

Siede, M. (2015). The plants of Phaenias. In *Phaenias of Eresus. Text, Translation, and Discussion*, Hellmann, O. and Mirhady, D. (eds). Transaction Publishers, New Brunswick/London, 351–364.

Touwaide, A. (1993). Strategie terapeutiche. In *Storia del pensiero medico occidentale, Volume I*, Grmek, M. (ed.). Antichità e Medioevo, Laterza, Rome/Bari, 133–149.

Työrinoja, R., Lehtinen, A.I., Føllesdal, D. (1990). *Knowledge and the Sciences in Medieval Philosophy*. Yliopistopaino, Helsinki.

Weisheipl, J.A. (1994). *S. Alberto Magno e le scienze*. Studio Domenicano, Bologna.

Zucker, A. (2015). Phainias and the naturalistic legacy of the Peripatos. In *Phaenias of Eresus. Text, Translation, and Discussion*, Hellmann, O. and Mirhady, D. (eds). Transaction Publishers, New Brunswick/London, 351–364.

12

The Arab World

Fabian KÄS
University of Cologne, Germany

In the middle of the 9th century, shortly after the zenith of the Abbasid caliphate, the Christian physician and translator Ḥunayn ibn Isḥāq (d. c. 260 AH/873 CE) compiled an epistle enumerating Galenic and pseudo-Galenic writings and their Arabic and Syriac versions. This unique bibliography contains no less than 179 titles, most of which had been translated into one of the oriental languages used among Baghdad's doctors at that time.

Ḥunayn's short treatise is the most renowned witness of the formative phase of Arabic medicine, which adopted virtually all Greek books available at that time in Byzantium and its former eastern provinces, as well as other texts written in Syriac, Middle Persian and Sanskrit. In particular, the Greek medical literature became the role model of Arab physicians at all times. Rearranging the scattered pieces of information given by authors such as Hippocrates, Galen, Rufus, Oribasius, Paul of Aegina and Dioscorides, they created monumental medical handbooks highly esteemed both in the Orient and Medieval Europe. The ancient humoral theory was never questioned in the classical epoch of Islam. Charges of a lack of critique and innovativeness often found in the older research literature are, however, unjustified. In the times before the development of modern microbiology, pharmaceutical chemistry, oncology, surgery and genetics and the means of medicine were limited

History of Life Sciences and Medicine 1,
coordinated by Lorenzo PERILLI. © ISTE Ltd 2026.

in many respects, and neither Greek, Arab nor Paracelsian physicians were able to overstep these boundaries[1].

12.1. Sources

Our most important sources for the history of medicine in the premodern Islamicate countries are, undeniably, hundreds of Arabic[2] works of various length[3], which have been preserved for today's reading. As they were not written for modern philologists, they leave many open questions, especially regarding the biographies of their authors. Ego-documents, such as Ḥunayn's invaluable epistle[4], are extremely

1 Because of its limited length, the present paper can only highlight the most important topics. An interested reader may consult the excellent article "Ṭibb" by Emilie Savage-Smith (2002) in *Encyclopaedia of Islam* for further reading. *New edition*, the concise handbook *Medieval Islamic medicine* by Pormann and Savage-Smith (2007), the survey by Manfred Ullmann (1978) entitled *Islamic medicine* and Johann Christoph Bürgel's (2016) study *Ärztliches Leben und Denken.* The most important bio-bibliographic studies are still volume III of Fuat Sezgin's (1970, English version 2023) *Geschichte des arabischen Schrifttums* and Ullmann's (1970) *Die Medizin im Islam.* Lucien Leclerc's (1876) *Histoire de la médecine arabe* is of course dated, but still worth reading. To minimize the bibliography, older standard editions already indicated in these secondary sources will not be listed. Instead, I will focus on new editions, especially translations into modern languages, making these texts accessible to nonspecialists as well.

2 The present paper focuses on writings in Arabic as the classical language of oriental medicine. Although Syriac was important in the first centuries of our period of investigation, only fragments of the numerous translations into this Aramaic language mentioned in Ḥunayn's epistle survived (Degen 1981; Kessel in Lamoreaux 2016, pp. 168–192). The Middle Persian medical books, some of which had been translated into Arabic, are almost completely lost. Modern Persian books are sporadically attested from the 4th/10th century on and became more and more important in the postclassical epoch (Elgood 1951). Turkish also became a language of scientific medicine at a rather late date (Shefer-Mossensohn 2009).

3 The bio-bibliographic sources sometimes even list miscellanies of one single page. The longest work was the partially preserved medical encyclopedia entitled *al-Shāmil* by Ibn al-Nafīs (d. 687/1288). The author was only able to finish 80 out of projected 300 volumes (Ullmann 1970, p. 173).

4 The text entitled *Risāla ilā ʿAlī ibn Yaḥyā fī dhikr mā turjima min kutub Jālīnūs bi-ʿilmihī wa-baʿḍ mā lam yutarjam* ("Epistle to ʿAlī ibn Yaḥyā mentioning the books by Galen which have – to his knowledge – been translated and some that have not yet been translated") is preserved in three divergent manuscripts. The latest edition (Lamoreaux 2016) does not contain the second part listing the texts deemed nonauthentic by Ḥunayn originally written as an independent "Treatise on books not mentioned by Galen in his list of books". Therefore, the additional consultation of the first editions by Bergsträsser (1925, 1932) is strongly advised. For the Arabic reception of Galen in general, see Bouras-Vallianatos and Zipser (2019, pp. 163–316) and for Ḥunayn's role, see Bouras-Vallianatos and Zipser (2019, pp. 179–195 Cooper).

rare, and most books only allow conclusions about the author's education, lifetime and patrons. Physicians were only rarely mentioned in standard chronicles, for example, as personal doctors of rulers occasionally involved in conspiracies. The rich Islamic prosopographic literature collecting information about religious scholars, judges, ascetics and others is only rarely useful for our context, since many physicians were Christians and Jews and only a few of the Muslims doctors had a theological education as well. Another problem is the fact that many texts written before c. 300/900 are lost and the scattered fragments ascribed to early authors do not allow reconstructions of their oeuvres.

Under these circumstances, a small number of bio-bibliographical works are of greatest interest in the history of early Arabic medicine. Because of the vast number of his own books and treatises, Galen himself had compiled lists of his own genuine works and the recommendable order of studying them (Ullmann 1970, pp. 35–37; Boudon-Millot 2007). This apparently motivated the translators Ḥunayn and Qusṭā ibn Lūqā[5] to write treatises pursuing the same objectives. Other important models for the first Arabic biographies of physicians were late antique texts gathering apophthegms and legendary stories of the lives of philosophers and doctors such as Galen and especially Hippocrates (Pinault 1992). The knowledge of Arabs about the lives of highly esteemed ancient authors was always restricted to the contents of these trivial legends and a few authentic pieces of information found in their own works. Drawing on such precursors and inspired by the Arabic literature on theologians, philologists and poets, the Andalusian Ibn Juljul (fl. 377/987, Kahl 2023) collected short biographies of ancient and oriental physicians up to his own time. Similar collections, including philosophers and scientists, are the *Ṭabaqāt al-umam* ("Classes of the nations") by his compatriot Ṣāʿid al-Andalusī (d. 462/1070, Blachère 1935) and Ibn al-Qifṭī's (d. 646/1248) *Taʾrīkh al-ḥukamāʾ* ("History of sages"). Ibn al-Nadīm's (d. 385/995) renowned "Index" (*al-Fihrist*) of books available in Baghdad is also useful for medical purposes (Dodge 1970). The most important source in our context is Ibn abī Uṣaybiʿa's (d. 668/1270) voluminous collection of biographies of physicians, including invaluable lists of their writings[6]. The vast medical literature written after Ibn abī Uṣaybiʿa's epoch-making book is

5 Ḥunayn's contemporary and fellow Christian Qusṭā ibn Lūqā (d. ca. 300/912) wrote a short treatise on the order of studying medical books especially for students. He recommends a number of books by Galen, Hippocrates, Oribasius, Paul and other Greek physicians. He apparently did not excerpt Galen's own list, but Ḥunayn's "Epistle" (Käs 2020).

6 This book bearing the title *ʿUyūn al-anbāʾ fī ṭabaqāt al-aṭibbāʾ* ("Sources of information regarding the classes of physicians") was recently edited with an annotated English translation in five volumes by a group of scholars (Savage-Smith et al. 2020).

still understudied. One reason for this fact is the scarcity of biographic information on authors living after the 8th/14th century[7].

12.2. Physicians

Before turning to the historical development of medicine in Islamicate countries, it may be useful to provide an outline of the persons active as physicians, their education and their professional ethics.

The biographical sources, especially Ibn abī Uṣaybiʿa's book, abound with anecdotes about the doctors, with a special focus on their relations with the court, their rivalries and curious patient records. Although these have not much historical value and many of them seem to be made up, they clearly show the reader how an ideal physician should behave, when he treats his patients, or when he is invited to court. In particular, prognostic skills and a certain instinct (*ḥads*) regarding the causes of diseases were highly praised (Bürgel 2016, p. 254). This ideal conception has, again, ancient precursors.

On the basis of the Greek biographical sketches of Hippocrates and Galen, the Arabs gave accounts of their ethics, professional skills, and habits, serving as role models for all other physicians (Pinault 1992, pp. 95–124; Bürgel 2016, pp. 400–408). Apart from such secondary sources, the works of the renowned Greeks also contain great amounts of deontological material.

It is a well-known fact that many physicians were non-Muslims in the first centuries after the Islamic expansion. The famous literary figure al-Jāḥiẓ (d. 255/869) told the funny story of a fictional Arab Muslim doctor called Asad ibn Jānī, who was workless, since all potential patients consulted Aramaic Christian doctors only[8]. The reason for this preference can easily be understood: al-Jāḥiẓ wrote his book at exactly the same time, when the bulk of the Greek and Syriac texts was translated into Arabic. Before this time, the classical medical literature was inaccessible to Arabs without knowledge of foreign languages. Later, many rulers used to choose Christians and Jews as their personal doctors since they were deemed to be less dangerous, as they were outsiders from politically active Islamic society. Famous examples are Saladin and his Ayyubid successors, who employed eminent non-Muslim experts, such as Maimonides (d. 600/1204) and Ibn Jumayʿ

7 Scattered pieces of information on later physicians excerpted mostly from chronicles and prosopographic sources were collected by ʿĪsā (1982; 1st edition 1942).

8 The story of Asad ibn Jānī is part of the "Book of misers" (*Kitāb al-Bukhalāʾ*, ed. ʿUmar al-Ṭabbāʿ, Dār al-Arqam, Beirut 1998, pp. 176f.; cf. Bürgel 2016, p. 223).

(d. 594/1198)[9]. The most remarkable family of Christian doctors is that of Jurjīs ibn Jibrīl ibn Bukhtīshūʿ (d. after 151/768), the former director of the hospital of Jundīshābūr in southern Iran. Descendants of the Bukhtīshūʿ family can be traced back to the 5th/11th century in Baghdad, where they served the caliphs and authored books (Ullmann 1970, pp. 108–111).

The medical profession was often inherited in families. Another famous example is that of Ḥunayn, whose son Isḥāq and nephew Ḥubaysh were translators as well. Many physicians received their basic education from their fathers or other relatives. We can also observe students gathered around an external master, such as Ibn Māsawayh (d. 243/857). Such intimate circles occasionally developed into veritable private schools, for example, the "Dakhwāriyya" in Damascus, which was founded by Muhadhdhab al-Dīn al-Dakhwār (d. 628/1230). Notably, medical education always had a non-state character, although it sometimes included practical training in the renowned hospitals of the Islamic capitals run by charitable foundations (Bürgel 2016, pp. 111–126).

In contrast, the examination of physicians, to protect the public from charlatans, belonged to the duties of the government. They were mostly controlled by market inspectors (*muḥtasib*). Especially in Baghdad, the "head of physicians" (*raʾīs al-aṭibbāʾ*) was entrusted with the examination of would-be doctors. The curriculum of students did not only include the Greek classics – especially the sixteen propaedeutic books by Galen – and the Arabic pandects but also collections of possible exam questions (Bürgel 2016, pp. 177–214). ʿAbd al-ʿAzīz al-Sulamī (d. 604/1207) compiled a study manual consisting of such questions and answers, including chapters dedicated to the skills of ophthalmologists, surgeons and bonesetters (Leiser and Al-Khaledy 2004). The most interesting book of this genre is certainly Ibn Buṭlān's (d. c. 458/1066) *Daʿwat al-aṭibbāʾ*, in which he presents medical knowledge in the form of amusing dialogs of physicians and a charlatan at a dinner party (Klein-Franke 1984a).

Deontology and professional ethics – regularly alluded to in the abovementioned anecdotes portraying ideal physicians – were only rarely treated in monographs (Ullmann 1970, pp. 223–227). One illustrious exception is Isḥāq ibn ʿAlī al-Ruhāwī's *Adab al-ṭabīb*. The book of this otherwise unknown Christian, who allegedly lived in Edessa at the end of the 3rd/9th century, builds on a popular

9 Maimonides wrote, among many other books, personal health tips for al-Malik al-Afḍal (Ullmann 1970, pp. 167–169). The texts have recently been edited and translated by Bos in the series "The medical works of Moses Maimonides". Ibn Jumayʿ authored an interesting epistle (Fähndrich 1983) to Ṣalāḥ al-Dīn, dedicated to the revival of the art of medicine, whose decline he deplored: a classical commonplace known from Galen on.

autochthonous Arabic genre dedicated to the skills and ideal conduct of professional groups such as secretaries or judges. It contains, among others, chapters on how the doctor should behave in any situation, the nobility of medical art; and the duties of the government[10]. Another thematically related monograph is the short protreptic treatise *al-Tashwīq al-ṭibbī* by one Ṣāʿid ibn al-Ḥasan (5th/11th century) (Bürgel 2016, pp. xxviiif.).

12.3. Alternative conceptions

One common feature of the texts mentioned in the preceding paragraph is the distinction of "true physicians" and quacks. The charlatans described are not only incompetent Hippocratic doctors but also healers without classical medical education. This professional disdain also included non-Arab "colleagues", such as a Frankish physician in the age of the crusades reportedly killing his patients with his barbarian therapies (Hitti 1964, p. 162). The focus of these polemics was on folk medicine and magic practices (Pormann and Savage-Smith 2007, pp. 145–161). As the former was transmitted orally, we cannot say much about it. However, the popular use of simple and compound drugs was closely related to that of scientific medicine[11].

Magical cures involving talismans, amulets, astrology or incantations were unworthy of discussion for serious physicians. It is an undeniable fact that premodern cultures, including the classical Greek and Islamicate Arabic civilizations, did not share our definition of "magic" common since the age of the Enlightenment. The struggle with therapies outside the system of humorism is deeply rooted in the Greco–Arabic scientific medicine. However, there was no clear distinction between conventional and magical practices. The therapeutic use of the body parts of animals and humans allegedly having "specific properties" (*khawāṣṣ*) was widely accepted (Raggetti 2018). "Iatromathematics", the medicinal implications of astrology, were disputed, especially since the Greek tradition was not free from such influences (Ullmann 1978, pp. 111–114; Klein-Franke 1984b). In the great pandects, we find detailed accounts of astrological topics only as late as in Dāwūd al-Anṭākī's (d. 1008/1599) *Tadhkirat ulī l-albāb* (Ullmann 1970, p. 181).

10 Levey (1967); Pormann and Savage-Smith (2007, p. 89); Bürgel (2016, pp. xxv–xxvii, 85). Levey's English translation, at that time based on the unique manuscript only, is, unfortunately, not free from mistakes.

11 We will see below that some physicians wrote special treatises for poor patients. The classical Greco–Arabic medicine was still popular among the common people in the 19th and 20th centuries, when the elites already trusted in modern Western medicine only.

The alternative conceptions addressed here also have one truly Islamic component, namely, the “Prophetic Medicine” (Pormann and Savage-Smith 2007, pp. 71–75, 145–161; Bürgel 2016, pp. 423–437). The second most important source of Islamic law, after the *Qur’ān*, are the sayings and deeds of Muḥammad transmitted by his early followers. The Prophet occasionally talked about nature and the healing of diseases as well (cf. T. Fahd, Ruḳya, *EI²* VIII 600). Needless to say, the opinions found in the collections of *ḥadīth* are in many respects different from those held by conventional physicians. The emergence of a literary genre dealing with Prophetic medicine is a relatively young phenomenon. Its classics by Ibn Qayyim al-Jawziyya or al-Dhahabī, a renowned historian, date from the 8th/14th century (Ullmann 1970, pp. 185–189).

A general question in the disputed area between religion and medicine is whether a pious patient should trust more in God’s decrees (*tawakkul*) or in the skills of physicians. A parable-like answer was given by Ibn Hindū (d. c. 422/1031), who relates the anecdote of an ascetic, who unsuccessfully tried to charm a venomous snake with Qur’ānic verses (Tibi 2010, p. 11; Bürgel 2016, pp. 20–34, esp. p. 29).

12.4. Historical development

The legendary origins of Arabic medicine are also related to the Prophet of Islam. One of his companions, a certain al-Ḥārith ibn Kalada (*EI²,* XII 354–355 Ch. Pellat), was allegedly a classically educated physician. Be that as it may, the treatises ascribed to him are certainly younger pseudepigraphs. The medical activities of another contemporary of Muḥammad, Ibn abī Rimtha, are also shrouded in legend. At the time of the conquest of the Persian kingdom and the eastern provinces of the Byzantine empire, Arabs encountered flourishing centers of medicine. The school of Alexandria[12] survived until c. 100/719 and that of Jundīshābūr, with its renowned hospital, was still active in the Abbasid era. Several Arabic works preserve fragments ascribed to one Tiyādhūq – perhaps Greek Θεοδόκος – and “the Jew” Māsarjawayh. Reliable biographical data from these two physicians are unavailable, but they may have been active around the 2nd/8th century (Ullmann 1970, pp. 19–24).

12 In the 1st/7th century, several Greek works by were written in Alexandria. These include texts by John the Grammarian, a priest called Ahrun (Aaron), and the abridgements of Galen’s classical propaedeutic works known as “Summaria Alexandrinorum” (*Jawāmi‘ al-Iskandarāniyyīn*). The transfer of the school via Antioch and Ḥarrān to Baghdad reconstructed from scattered notes by Arab historians is, according to modern research, just a legend.

12.4.1. *Translations*

As we have already seen at the beginning, the real formative period of Arabic medicine starts with translations of the 3rd/9th century. Because of their language skills and since medical and philosophical school traditions had also survived in Syria and Iraq, Aramaic-speaking Christians had a leading role in this translation movement.

Within a few decades, most of the medical – and of course also philosophical and scientific – texts available at that time in the Orient and in Byzantium were translated into Arabic. Importantly, the most substantial losses of ancient scientific literature had taken place long before. The oriental tradition certainly preserved some works no longer available in Greek today. However, the bulk of the Hellenistic literature predating Galen was already lost in antiquity.

The translators into Arabic often drew on Syriac intermediary versions, occasionally authored centuries earlier, for example, by Sergius of Rēsh ʿAynā (d. 536). We learned from Ḥunayn that many books had been translated more than once, especially since the quality of such versions was not always the best. Another important conclusion drawn from his "Epistle" is that the translations were financed by private commissioners, often physicians not mastering Greek, such as Ibn Māsawayh. This clearly proves that older theories of state sponsorship by the Abbasid government are wrong. Some caliphs, for example, al-Maʾmūn (d. 218/833), were highly interested in science and perhaps fostered this movement in some form or another, but they certainly did not run a translation office in the "house of wisdom" (*bayt al-ḥikma*), which may simply have been a palace library (Gutas 1998; Pormann and Savage-Smith 2007, pp. 24–29).

Syriac was not the only intermediary language[13]. Some Greek texts had been first translated into Middle Persian, such as the book on substitute drugs by Badīghūras (pseudo-Pythagoras). Indian books on medicine, which were highly esteemed in the first centuries, were mostly transferred into Arabic via Pahlavi. Notably, the Arabs also came into contact with autochthonous Persian and Syriac texts (Kahl 2015). The latter were especially important since comprehensive handbooks became the model of Arabic pandects referred to as *kunnāsh*, an Aramaic loanword.

---

13 A few Latin historical and scientific texts have been translated in al-Andalus into Arabic as well. This field can be neglected here, since we have no such books dealing with medicine. It cannot, however, be excluded that some texts – perhaps a version of one of the handbooks by Oribasius – had indeed been translated from Latin (Ibn Janāḥ 2020, p. 124).

12.4.2. *Classical epoch and European reception*

The best proof of the high esteem of Arabic authors for the Greeks is the fact that even those who did not indicate sources of their statements[14] occasionally cited Galen, Hippocrates or Paul to stress their own classical education. The geographer and pharmacologist al-Idrīsī even referred to his copy of Dioscorides's book on simple drugs as his "*muṣḥaf*": a term normally reserved for the *Qur'ān* (Käs 2010, p. 125). In the first centuries, the medical training also focused on the classics, especially Hippocrates' main works and Galen's "sixteen books" (Bürgel 2016, pp. 140–162). Later, textbooks that were easier to use, such as Ibn Sīnā's "Canon of Medicine", became increasingly popular.

As stated above, the theory of humorism was not challenged by Arabs before the 10th/16th century. This does, however, not mean that they always followed the Greeks uncritically. In particular, Galen's philosophical ideas had been questioned as early as antiquity. Arabic authors, such as al-Rāzī (d. c. 313/925) or Ibn Rushd (d. 595/1198), continued this critique and casted doubt on some of his medical concepts (Ullmann 1970, pp. 67f.; Koetschet 2019). The means of premodern medicine were limited in many respects, which is why the Arabs were unable to add absolutely new findings overlooked by the Greeks. Noteworthy exceptions are ʿAbd al-Laṭīf's and Ibn al-Nafīs' contributions to anatomy (see below).

During the classical epoch of Islamicate medicine between the 3rd/9th and the 7th/13th centuries, many monographs and handbooks were produced by many authors (cf. Sezgin 1970). The outline of the medical literature given below will focus on writings from this era. These centuries were, of course, the time of the culmination of the oriental culture as well. The decline after the middle of the 13th century had mostly political reasons, such as the fall of Baghdad and the loss of al-Andalus as a result of the Reconquista. The medical achievements of Arabs in this epoch can hardly be characterized in a few ways. The most important of them was, perhaps, a hitherto unknown degree of systematic clarity. Even Galen himself had to confess that his oeuvre was difficult to view. Arabic handbooks and vade mecums can, by contrast, be easily utilized, some even by laymen. Needless to say, the amount of Arabic literature preserved until today exceeds, despite some losses, that of ancient medical works by far.

14 In the Arabic tradition, we can distinguish between two approaches of citations. Some authors meticulously mentioned the sources of each piece of information. In the end, even voluminous works represent mere collections of quotations and fragments exactly ascribed to specific authors and books. The physicians from al-Andalus were unrivalled in exactitude of their citations (Käs 2010, p. 2). Authors from Kairouan, as well as those living in the Islamicate East from the 4th/10th century onward, normally avoided mentioning names.

The undisputed center of early Arabic medicine was the Abbasid capital of Baghdad, which also attracted physicians from Iran, such as Ibn Rabban al-Ṭabarī and al-Rāzī. From the 4th/10th century onward, this science became increasingly multipolar, which was certainly also an effect of the decline in the caliphs' power. Especially important were larger Iranian towns and residential cities, where we find authors such as al-Bīrūnī and Ibn Sīnā. Damascus's influence grew relatively late, after the foundation of the Nūrī hospital[15] and al-Dakhwār's school of medicine. The North African center was not Cairo but Kairouan, which housed in the 4th/10th century a school of renowned physicians, including Isḥāq ibn ʿImrān, Isḥāq al-Isrāʾīlī, and Ibn al-Jazzār. Of greatest importance was the school of al-Andalus, with members such as the historian Ibn Juljul, the pharmacologists Ibn Janāḥ and Ibn al-Bayṭār, as well as the authors of handbooks al-Zahrāwī and Ibn Zuhr.

The relatively easy accessibility of Arabic medical books and their clear systematization were the main reasons for their medieval European reception. Their Latin translations were highly esteemed and served for centuries as textbooks for the education of physicians. The first renowned translator was Constantine the African, who hailed from Tunis. He settled in Italy in the monastery of Montecassino, where he died in 1087. He was well acquainted with the literature of the Kairouan school, which he translated. His oeuvre includes Ibn ʿImrān's book on melancholy, Isḥāq al-Isrāʾīlī's texts on foodstuffs and uroscopy, the handbook *Viaticum peregrinantis* by Ibn al-Jazzār, as well as the pandect entitled *al-Malakī* by al-Majūsī (Schipperges 1964, pp. 17–54). The Latin translation movement lasted for approximately 200 years with centers in Toledo, Chartres and Southern Italy, especially at the court of Emperor Frederick II. The most famous translator was certainly Gerard of Cremona (d. 1187), who authored the Latin versions of Ibn Sīnā's *Canon medicinae*, al-Zahrāwī's *Chirurgia* and al-Rāzī's *al-Manṣūrī*. The latter's *al-Ḥāwī* (*Continens*) was translated a century later by Faraj ibn Sālim (Schipperges 1964, pp. 92–96). Important intermediaries of Arabic knowledge in Europe were also Jewish translators with their Hebrew versions of these books (Steinschneider 1893).

12.4.3. *Postclassical medicine*

When Ibn abī Uṣaybiʿa wrote his collection of biographies in the mid-7th/13th century, the Islamic world had changed in many respects. The caliphate of Bagdad no longer existed, Egypt was ruled by rivaling Mamluks and Spain was almost entirely in the hands of the Christians. These changes also had an impact on the

15 For hospitals in Islamic capitals, see the article Bīmāristān, In *EI*², I 1222–1226 (Colin and Şehsuvaroğlu).

Arabic scientific and medical literature. Despite the historical decline, many books on medicine were written in the following centuries, but they were different from those of the preceding epoch. This era abounds with commentaries and abridgements of the classics, for example, Ibn Sīnā's Canon, or, in the field of pharmacognosy, Ibn al-Bayṭār's book on simple drugs. Genuine new textbooks, such as *al-Ḥāwī fī ʿilm al-tadāwī* of al-Shīrāzī (d. 730/1330), are rare and based on older pandects. A last classical comprehensive handbook was Dāwūd al-Anṭākī's *Tadhkira*, which includes unusual content, such as astrology. The epoch after 700/1300 also witnessed the emergence of the literature on Prophetic medicine. Closely related to the latter are books on epidemics and the plague, which were increasingly dominated by religious contents (Ullmann 1970, pp. 246–250).

The Ottoman court attracted the best physicians of this time (Shefer-Mossensohn 2009). It was also in Constantinople, where the first contacts with contemporaneous European medicine took place. Since the 16th century, the time-honored system of humorism had been challenged by Paracelsus (d. 1541) and his followers, such as Oswald Crollius (d. 1608) and Daniel Sennert (d. 1637). Their books on "new chemical medicine" were also translated into Arabic. They were intensively received by the Ottoman court physician Ibn Sallūm (d. 1080/1669), who wrote his own books on this concept (Bachour 2012). Greco–Arabic medicine was never completely replaced by modern Western medicine. Especially on the Indian subcontinent, where it is referred to as the "Unani Tibb" (i.e. Greek medicine), it is still practiced alongside Ayurveda (Pormann and Savage-Smith 2007, pp. 173–179).

12.5. Disciplines and literary genres

In what follows, I provide a brief outline of the disciplines of medicine and their literary genres[16]. It must be stressed that medical specialists do not exist in the modern sense. Only surgeons and ophthalmologists form professional groups differentiated from general practitioners (Leiser and Al-Khaledy 2004, pp. 89–110).

12.5.1. *Pandects*

Currently, the most renowned Arabic books on medicine are a group of comprehensive handbooks covering "the art" as a whole. Some of these "pandects" had a great impact on medieval and early modern medicine in Europe via Latin, Hebrew and Byzantine Greek translations. The Arabic technical term for such a handbook is *kunnāsh*, a Syriac loanword, which reveals the origins of this genre.

16 Further information on the authors mentioned in this passage can easily be found in Sezgin (1970), Ullmann (1970), or in articles dedicated to them in *EI*[2].

The tradition even dates back to late antiquity and handbooks by Oribasius, Paul of Aegina or Alexander of Tralles. The oldest preserved texts of this genre are the fragmentary *kunnāsh* by Ibn Sarābiyūn (d. c. 250/864) – reportedly a translation from Syriac – and Ibn Rabban al-Ṭabarī's (d. after 240/855) *Firdaws al-ḥikma* ("Paradise of Wisdom"). Abū Bakr al-Rāzī (d. c. 313/925) is famous for two handbooks, which were both highly esteemed in the Orient, namely, the concise *al-Manṣūrī* (*Liber ad Almansorem*) and the comprehensive *al-Ḥāwī* (*Continens*), preserving large amounts of quotations from older texts. ʿAlī ibn al-ʿAbbās al-Majūsī's *al-Malakī* (4th/10th century) was translated twice into Latin: as *Pantegni* by Constantine and as *Liber regius* by Stephen of Antioch in 1127. The most renowned Arabic textbook is, undoubtedly, the *Qānūn fī l-ṭibb* by Abū ʿAlī ibn Sīnā (d. 428/1037), the famous philosopher known in Europe as Avicenna. Pandects written in al-Andalus are almost completely lost (Janāḥ 2020, pp. 121–127, 154). Noteworthy exceptions are al-Zahrāwī's *al-Taṣrīf* (see below) and Ibn Zuhr's (d. 557/1162) *al-Taysīr*.

The voluminous pandects were often difficult to use, even for specialists. This led to the emergence of abridgements and therapeutic manuals, such as Ibn al-Jazzār's (4th/10th century) *Zād al-musāfir* ("Provisions for the traveler"). Special forms of concise presentations of medical knowledge include branch diagrams (*tashjīr*) and tables (*taqwīm*, see below). An early example of the former is Ibn Māsawayh's (d. 243/857) *al-Mushajjar*. Books in the form of question and answer were also popular, such as Ḥunayn's "Medical questions" (*Masāʾil fī l-ṭibb*).

12.5.2. *Anatomy and nosology*

Anatomy played an important role in medical education, which is why most *kunnāshs* begin with this topic (Savage-Smith 2007). The ancient Greek knowledge about the human body could, however, be surpassed in only a few respects. ʿAbd al-Laṭīf (d. 629/1231) found out that the human mandible consists of only one bone, and Ibn al-Nafīs reported pulmonary blood circulation (Ullmann 1970, pp. 171, 174). Since the latter's works were unknown in the West, this highly important finding, which could have initiated a medical revolution, remained unnoticed for centuries.

Ancient concepts about health and disease were also passed on without large-scale changes (Ullmann 1978, pp. 55–69). As we have seen, the system of humorism was undisputed in premodern times. An interesting case is that of contagion. The Greeks had discarded this theory, since epidemic diseases were caused, in their view, by poisonous vapors, so-called "miasmata". Although Muslims mostly accepted this, there are a few *ḥadīths* of the Prophet supporting the idea of contagious diseases. This has led to long controversies, especially in the literature on pestilential diseases (Ullmann 1978, pp. 86–96; Pormann and Savage-Smith 2007, p. 58f.).

12.5.3. *Dietetics and pharmacology*

For Greco–Arabic physicians, the preservation of health was in many respects more important than the treatment of diseases having already occurred. One of the fundaments of a correct way of living is adequate nutrition. The Arabic literature abounds with books giving such health tips, sometimes even for a single patient. One interesting specimen is the book on dietetics in the form of tables written by Ibn Buṭlān, the *Taqwīm al-ṣiḥḥa*[17].

Equally important was Isḥāq al-Isrāʾīlī's (d. c. 320/932) book on foodstuffs (*K. al-Aghdhiya*). Dietetics are, of course, also present in the pandects as well as in books on pharmacognosy, as there was no clear differentiation between drugs and foods.

The Arabic literature on simple drugs is extremely rich and flourished until the 12th/18th century (Käs 2010, pp. 1–197). It is based on Greek precursors and especially Dioscorides's *De materia medica* translated at least three times into Arabic. The culmination of the Arabic pharmacognostic literature is Ibn al-Bayṭār's (d. 646/1248) *al-Jāmiʿ* ("the Collector") gathering information on hundreds of drugs (Leclerc 1877–1883), many of which had not been mentioned by ancient physicians, such as myrobalans, tamarinds, tabasheer and betel nuts.

Officinal drugs also include minerals (Käs 2010) and animal products – either mentioned as foodstuffs or remedies with sympathetic properties (Raggetti 2018). The focus of this literature is on plants, and their fluent nomenclature led to the emergence of a subgenre dedicated to the explanation of phytonyms[18]. Drugs were mainly used in the form of compounds, such as pills, decoctions, salves and powders. Pharmaceutic chapters can regularly be found in pandects, and there are lots of monographic dispensatories (Kahl 2007).

---

17 Manuscripts of the Latin translation entitled *Tacuinum sanitatis* were often luxuriously illustrated for princely readers (Ullmann 1978, p. 99).

18 Many medicinal plants and vegetables have no autochthonous Arabic names, since they do not grow on the Peninsula. Instead, the Arabs adopted Persian, Indian and Greek phytonyms. In the Islamicate West, where this literature flourished, the authors were in addition confronted with Romance and Berber terms (Ibn Janāḥ 2020).

12.5.4. *Surgery and ophthalmology*

We learned from the Hippocratic oath that practical surgery was not considered part of medical art in a strict sense by Greco–Arabic physicians. Surgeons, bonesetters and bloodletters formed, such as druggists, own professional groups. Descriptions of interventions appear in the *kunnāshs*, especially in the form of quotations from ancient sources. There are only a very small number of Arabic medical texts dealing with surgery (Pormann and Savage-Smith 2007, pp. 121–129). The most important of them are the monograph *al-ʿUmda fī l-jirāḥa* ("the pillar of surgery") by Ibn al-Quff (d. 685/1286) and the exceptional thirtieth chapter of Abū l-Qāsim al-Zahrāwī's (5th/11th century) handbook referred to in Latin as *Chirugia Albucasis*. The latter was very influential in the West until the 18th century, since al-Zahrāwī had exceeded the contents of his sources, especially Paul of Aegina, by far (Spink and Lewis 1973).

The situation of ophthalmic surgery, especially trachoma and cataract interventions, is not very different. Only the number of literary sources is higher, because many authors on ophthalmology describe such procedures (Pormann and Savage-Smith 2007, pp. 125–128). The history of medicine of the eye among Arabs is particularly interesting since most Greek texts dedicated to this genre are lost. In contrast, many Arabic treatises written from the 3rd/9th century on, for example, by Ḥunayn and Ibn Māsawayh, are still preserved (Ullmann 1970, pp. 204–214).

12.5.5. *Special groups of patients*

Although gynecology was practiced not by male doctors but by midwives, the medical literature often covers related topics. Al-Zahrāwī's surgery contains detailed descriptions of the specula and instruments for extracting dead fetuses (Spink and Lewis 1973, pp. 484–501). Anatomy and diseases of the female genitals are regularly found in handbooks. Monographs on gynecology and obstetrics are indeed almost completely missing, except for a remarkable book by ʿArīb ibn Saʿīd (4th/10th century) (Ullmann 1970, p. 139). It includes passages on pediatrics, which were also the topic of a monograph by al-Rāzī (Bos and McVaugh 2015). The Arabic medical literature abounds with health tips for special groups of persons. Ibn al-Jazzār wrote, for example, in addition to his therapeutic manual, regimens for children, elderly people, poor patients and persons suffering from obliviousness (Sezgin 1970, pp. 306f.).

12.6. Conclusion

This chapter may have shown that there was no hiatus between ancient Greek and medieval Arabic medicine. Muslims conquered some of the richest provinces of the Roman and Persian empires. The medical traditions they encountered there continued without a break. Even ancient magic and superstition were inherited by Arabs. Despite this continuity, we can also observe specific differences. Some of them are based on the fact that the Islamicate world is far larger than the Mediterranean area of Greco–Roman antiquity. Persian and Indian traditions as well as new medicinal drugs enriched Arabic medicine. Another source of differences is indeed the religion of Islam, as we have seen in the context of gynecology, Prophetic medicine, discussions about contagion and the emergence of hospitals run by religious endowments.

12.7. References

Bachour, N. (2012). *Oswaldus Crollius und Daniel Sennert im frühneuzeitlichen Istanbul.* Centaurus, Freiburg.

Bergsträsser, G. (1925). *Ḥunain ibn Isḥāq über die syrischen und arabischen Galen-Übersetzungen. Zum ersten Mal herausgegeben und übersetzt.* F.A. Brockhaus, Leipzig.

Bergsträsser, G. (1932). *Neue Materialien zu Ḥunain ibn Isḥāq's Galen-Bibliographie.* F.A. Brockhaus, Leipzig.

Blachère, R. (1935). *Ṣâ'id al-Adalusî. Kitâb Ṭabaḳât al-umam (Livre des catégories des nations).* Larose Éditeurs, Paris.

Bos, G. and McVaugh, M. (2015). *Al-Rāzī, On the Treatment of Small Children (De curis puerorum). The Latin and Hebrew Translations.* Brill, Leiden.

Boudon-Millot, V. (2007). *Galien. Tome I. Introduction générale. Sur l'ordre de ses propres livres. Sur ses propres livres. Que l'excellent médecin est aussi philosophe.* Les Belles Lettres, Paris.

Bouras-Vallianatos, P. and Zipser, B. (2019). *Brill's Companion to the Reception of Galen.* Brill, Leiden.

Bürgel, J.C. (2016). *Ärztliches Leben und Denken im arabischen Mittelalter. Bearbeitet von F. Käs.* Brill, Leiden

Degen, R. (1981). Galen im Syrischen: Eine Übersicht über die syrische Überlieferung der Werke Galens. In *Galen: Problems and Prospects*, Nutton, V. (ed.). Wellcome Institute, London, 131–166.

Dodge, B. (1970). *The Fihrist of al-Nadīm. A Tenth-century Survey of Muslim Culture.* Columbia University Press, New York.

Elgood, C. (1951). *A Medical History of Persia and the Eastern Caliphate*. Cambridge University Press, Cambridge.

Fähndrich, H. (1983). *Treatise to Ṣalāḥ ad-Dīn on the Revival of the Art of Medicine by Ibn Jumay'*. Steiner, Wiesbaden.

Gibb, H.A.R., Kramers, J.H., Lewis, B., Pellat, Ch., Schacht, J. (1954–2002). *The Encyclopaedia of Islam. New Edition*. Brill, Leiden.

Gutas, D. (1998). *Greek Thought, Arabic Culture. The Graeco-Arabic Translation Movement in Baghdad and Early Abbasid Society (2nd–4th/8th–10th Centuries)*. Routledge, London.

Hitti, P. (1964 [1929]). *Memoirs of an Arab-Syrian Gentleman or An Arab Knight in the Crusades. Memoirs of Usāmah Ibn-Munqidh*. Khayats, Beirut.

'Īsā, A. (1982 [1942]). *Mu'ǧam al-aṭibbā'. Ḏayl 'uyūn al-anbā' fī ṭabaqāt al-aṭibbā' li Ibn abī Uṣaybi'a*, 2nd edition. Dār al-Rā'id al-'Arabī, Beirut.

Janāḥ, Ibn (2020). *Marwān ibn Janāḥ: On the Nomenclature of Medicinal Drugs (Kitāb al-Talkhīṣ)*. Brill, Leiden.

Kahl, O. (2007). *The Dispensatory of Ibn al-Tilmīḏ*. Brill, Leiden.

Kahl, O. (2015). *The Sanskrit, Syriac and Persian Sources in the Comprehensive Book of Rhazes*. Brill, Leiden.

Kahl, O. (2023). *Ibn Juljul's Generations of Physicians and Sages. Translated from Arabic with Introduction and Indices*. Brill, Leiden.

Käs, F. (2010). *Die Mineralien in der arabischen Pharmakognosie*. Harrassowitz, Wiesbaden.

Käs, F. (2020). Qusṭā ibn Lūqā: Über die Reihenfolge der Lektüre medizinischer Bücher. *Galenos. Rivista di filologia dei testi medici antichi*, 14, 231–255.

Klein-Franke, F. (1984a). *Ibn Buṭlān. Das Ärztebankett*. Hippokrates-Verlag, Stuttgart.

Klein-Franke, F. (1984b). *Iatromathematics in Islam. A study on Yūḥannā Ibn aṣ-Ṣalt's book on "Astrological medicine"*. Olms, Hildesheim.

Koetschet, P. (2019). *Abū Bakr al-Rāzī, Doutes sur Galien*. De Gruyter, Berlin/Boston.

Lamoreaux, J.C. (2016). *Ḥunayn ibn Isḥāq on his Galen translations*. Brigham Young University Press, Provo.

Leclerc, L. (1876). *Histoire de la médecine arabe*. Imprimerie Nationale, Paris.

Leclerc, L. (1877–1883). *Traité des simples par Ibn el-Beïthar*. Imprimerie Nationale, Paris.

Leiser, G. and Al-Khaledy, N. (2004). *Questions and Answers for Physicians. A Medieval Arabic Study Manual by 'Abd al-'Azīz al-Sulamī*. Brill, Leiden.

Levey, M. (1966). *The Medical Formulary or Aqrābādhīn of al-Kindī Translated with a Study of Its Materia Medica*. University of Wisconsin Press, Madison.

Levey, M. (1967). Medical ethics of medieval Islam with special reference to Al-Ruhāwī's "Practical Ethics of the Physician". *Transactions of the American Philosophical Society. New Series. American Philosophical Society*, 57(3), 1–100.

al-Nadīm, Ibn (2009–2014). *K. al-Fihrist*. Al-Furqān Islamic Heritage Foundation, London.

Pinault, J.R. (1992). *Hippocratic Lives and Legends*. Brill, Leiden.

Pormann, P.E. (2004). *The Oriental Tradition of Paul of Aegina's Pragmateia*. Brill, Leiden.

Pormann, P.E. and Savage-Smith, E. (2007). *Medieval Islamic medicine*. Edinburgh University Press, Edinburgh.

Raggetti, L. (2018). *ʿĪsā ibn ʿAlī's Book on the Useful Properties of Animal Parts. Edition, Translation and Study of a Fluid Tradition*. De Gruyter, Berlin/Boston.

Savage-Smith, E. (2000). Ṭibb. In *Encyclopaedia of Islam. New Edition*, Bosworth, C.E. (ed.). Brill, Leiden, 10, 452–460.

Savage-Smith, E. (2007). Anatomy. In *Encyclopaedia of Islam. Three*, Fleet, K., Krämer, G., Matringe, D., Nawas, J., Stewart, D.J. (eds). Brill, Leiden, 126–131.

Savage-Smith, E., Swain, S., van Gelder, G.J. (2020). *A Literary History of Medicine. The ʿUyūn al-anbāʾ fī ṭabaqāt al-aṭibbāʾ of Ibn Abī Uṣaybiʿah*. Brill, Leiden.

Schipperges, H. (1964). *Die Assimilation der arabischen Medizin durch das lateinische Mittelalter*. Steiner, Wiesbaden.

Sezgin, F. (1970). *Geschichte des arabischen Schrifttums. Band III. Medizin, Pharmazie, Zoologie, Tierheilkunde*. Brill, Leiden.

Shefer-Mossensohn, M. (2009). *Ottoman Medicine. Healing and Medical Institutions, 1500–1700*. State University of New York Press, Albany.

Spink, M.S. and Lewis, G.L. (1973). *Albucasis. On Surgery and Instruments. A Definitive Edition of the Arabic Text with English Translation and Commentary*. Wellcome Institute, London.

Steinschneider, M. (1893). *Die hebräischen Übersetzungen des Mittelalters und die Juden als Dolmetscher*. Bibliogr. Bureau, Berlin.

Tibi, A. (2010). *Abū al-Faraj ʿAlī ibn al-Ḥusayn ibn Hindū: The Key to Medicine and a Guide for Students*. Garnet Publishing, Reading.

Ullmann, M. (1970). *Die Medizin im Islam*. Brill, Leiden.

Ullmann, M. (1978). *Islamic Medicine*. Edinburgh University Press, Edinburgh.

List of Authors

Tommaso ALPINA
Università di Pavia
Italy
and
American University of Beirut
Lebanon

Paola COSMACINI
Independent Researcher
Rome
Italy

Jean-Claude DUPONT
CHSSC
Université de Picardie Jules Verne
Amiens
France

Andrea FALCON
Concordia University
Montreal
Canada
and
Università degli Studi di Milano
Milan
Italy

Markham J. GELLER
University College London
UK

Fabian KÄS
University of Cologne
Germany

Vito LORUSSO
Liceo classico F. De Sanctis
Sant'Angelo dei Lombardi
and
Università degli Studi di
Napoli L'Orientale
Naples
Italy

Vivian NUTTON
University College London
UK

Christian ORSENIGO
Museo Civico di Crema
e del Cremasco
Italy

Lorenzo PERILLI
Università degli Studi di
Roma Tor Vergata
Rome
Italy

Luciana REPICI
Università di Torino
Turin
Italy

Amneris ROSELLI
Università degli Studi di
Napoli L'Orientale
Naples
Italy

Laurence TOTELIN
Cardiff University
UK

Dolly YANG
Independent Researcher
London
UK

Kenneth G. ZYSK
University of Copenhagen
Denmark

Index of Names

A

Index by Ambra Serangeli, University of Rome Tor Vergata.

H

I

J

K, L

M

N

O

P

Q, R

S

T, U

V, W

X, Y

Z

Index of Terms

Index by Francesco Botti, Camilla Lisotti, Ambra Serangeli, University of Rome Tor Vergata.

Q, R

S

T

U, V

W, X

Y, Z

Summary of Volume 2

Chapter 4. The Visible and the Invisible: Images of the Body, From Descartes to the Encyclopedists
Paolo QUINTILI

Chapter 5. The Observation and Classification of Living Beings: The Animal, 16th–18th Centuries
Jean-Luc GUICHET

Chapter 8. The Emergence of Transformism in the 18th and 19th Centuries

Stéphane TIRARD

Summary of Volume 3

Chapter 2. The Reception of Evolutionary Theory in France

Olivier PERRU

Chapter 3. The History of the Ecosystem Concept

Julien DELORD

Chapter 4. The Cell: The Basic Structural and Functional Unit of Living Organisms

François DUCHESNEAU

Chapter 5. The History of the Metabolism Concept: From Its Beginnings to Maturity

Jean-Claude DUPONT

Chapter 6. The History of the Gene

Antonine NICOGLOU

Chapter 9. Human Brains: Representations, Explorations and Stimulations
Mathilde LANCELOT

Chapter 10. Modern Medicine and the Sciences: A History of Ruptures
Mathieu ARMINJON

Printed and bound by CPI Group (UK) Ltd, Croydon, CR0 4YY

13/09/2026

14961072-0008